AS/A-LEVEL YEARS 1 AND 2

STUDENT GUIDE

EDEXCEL

Physics

Practical assessment

Carol Davenport, Graham George and
Kevin Lawrence

Series editor: Graham George

HODDER
EDUCATION
AN HACHETTE UK COMPANY

Hodder Education, an Hachette UK company, Blenheim Court, George Street, Banbury, Oxfordshire OX16 5BH

Orders

Bookpoint Ltd, 130 Park Drive, Milton Park, Abingdon, Oxfordshire OX14 4SE

tel: 01235 827827

fax: 01235 400401

e-mail: education@bookpoint.co.uk

Lines are open 9.00 a.m.–5.00 p.m., Monday to Saturday, with a 24-hour message answering service. You can also order through the Hodder Education website: www.hoddereducation.co.uk

ISBN 978-1-4718-8570-9

First printed 2017

Impression number 5 4 3 2 1

Year 2020 2019 2018 2017

This guide has been written specifically to support students preparing for the Edexcel A-level Physics examinations. The content has been neither approved nor endorsed by Edexcel and remains the sole responsibility of the authors.

Cover photo: Peter Hermes Furian/Fotolia; page 71 GIPhotostock/SPL.

Typeset by Integra Software Services Pvt. Ltd, Pondicherry, India

Printed in Slovenia

Hachette UK's policy is to use papers that are natural, renewable and recyclable products and made from wood grown in sustainable forests. The logging and manufacturing processes are expected to conform to the environmental regulations of the country of origin.

Contents

Content Guidance

Questions & Answers

■ About this book

This guide is one of a series covering the Edexcel specification for AS and A-level physics. It offers advice for the effective development of practical skills. Its aim is to help you understand and put into practice the physics — it is not intended as a shopping list, enabling you to cram for the examination. The guide has four sections:

- The **Maths and Units** section incorporates the mathematical skills that you need for the AS and A-level examination, together with a reminder of the importance of units. For A-level you need more advanced mathematical skills, which are highlighted in **bold**. This section provides the basis of all calculations that you will be required to undertake and that is why it is the first section in the book. Without these mathematical skills, you will not be able to analyse your data and subsequently draw conclusions about your experiment. You are advised to work through this section to remind yourself of the essential mathematics and put your knowledge to the test by attempting the questions as you go along. You may also find it useful to refer back to this section when you are trying questions later in the book.

- The **Practical Skills** section develops the essential skills needed for successful practical work, based on the practical criteria set out in the specification. It will help you become a confident practical physicist, able to use a variety of apparatus and techniques to collect data, which can then be analysed and used to draw valid conclusions.

- The **Core Practicals** section reflects the Edexcel requirement that you should acquire competence and confidence in a variety of practical, mathematical and problem-solving skills and in handling apparatus competently and safely. Each of the 16 core practicals is discussed in detail, with worked examples based on experimental data. For the AS examination, you need only be familiar with core practicals 1–8.

- The **Questions & Answers** section pulls together the other three sections through a range of practice questions based on the type of question that will be asked in examination papers to test practical knowledge, particularly in A-level paper 3. It does *not* in itself replicate a complete examination paper, as it contains only questions that test practical skills. Answers are provided and common errors made by students are also highlighted, so that you, hopefully, do not make the same mistakes.

If you are reading this, you are clearly determined to do well in your examinations! If you try all the knowledge check questions, the worked examples and the questions at the end *before* looking at the answers, you will begin to think for yourself and develop the necessary techniques for answering examination questions. As the answers to the worked examples are an integral part of the learning process, they appear immediately after the questions. You are recommended to cover up the answers to prevent 'cheating' — after all, if you 'cheat' by looking up the answers first, you are only cheating yourself!

Thus prepared, you will be able to approach the examination with confidence. Good luck, and remember that physics is fun!

Content Guidance

■ Maths and units

Throughout your AS or A-level course, you will make use of a range of mathematical knowledge and techniques. Some of these, such as basic algebra, will be used regularly in many different topics. Others, such as logarithms, will only be needed in a few topics. In this section, we shall cover the range of mathematical skills required during your course, with a particular focus on how they are used in practical work. Those statements in **bold** will be assessed in the full A-level course only.

Arithmetic and units

Correctly interpreting numerical information is an important first step in many practical calculations in physics.

Throughout your studies you will need to:

- recognise and make use of appropriate units in calculations
- recognise and use expressions in decimal and standard form
- use ratios, fractions and percentages
- estimate results
- use calculators to find and use power functions, **exponential** and **logarithmic functions**
- use calculators to handle $\sin x$, $\cos x$ and $\tan x$ when x is expressed in degrees or radians

Physical measurements are expressed in standard (SI) units (Table 1). These have been agreed internationally so that the same value and unit for each is used in different countries.

Table 1 Standard SI base quantities and units

Quantity	Unit	Symbol
length	metre	m
mass	kilogram	kg
time	second	s
electric current	ampere	A
temperature	kelvin	K
amount of substance	mole	mol

These base SI units are combined to give units for other quantities. The units of many commonly used quantities are given their own names. For example, force is measured in newtons (N) but this can be expressed as $kg\,m\,s^{-2}$; pressure is measured in pascals (Pa), which is the same as $kg\,m^{-1}\,s^{-2}$.

Exam tip

Always look carefully at the units in any numerical data. It is often helpful to convert units into standard units such as metres (m), rather than leave them as millimetres (mm).

Knowledge check 1

Give the following units in terms of base SI units:
a joule, J
b coulomb, C
c ohm, Ω
d watt, W

For very small or very large measurements, prefixes are used. Table 2 gives some of the common prefixes used in physics.

Table 2 Standard SI prefixes

Name and symbol	Factor of 10	Name and symbol	Factor of 10
centi, c	10^{-2}		
milli, m	10^{-3}	kilo, k	10^{3}
micro, μ	10^{-6}	mega, M	10^{6}
nano, n	10^{-9}	giga, G	10^{9}
pico, p	10^{-12}	tera, T	10^{12}
femto, f	10^{-15}		

Knowledge check 2

A student is investigating the density of a length of wooden dowelling. She measures the diameter of the wood to be 12.7 mm and the length of the dowelling as 80.0 cm. She calculates that the area of cross-section of the dowelling is 127 mm^2 and that its volume is 102 cm^3. Convert these four measurements into m, m^2 and m^3, expressing your answers in standard form if appropriate.

Worked example

(**A-level only**) A student is measuring the gravitational attraction between two masses, M and m. Estimate what happens to the gravitational force between the masses when the distance separating them is halved.

Answer

For the original force:

$$F_s = \frac{GMm}{r^2}$$

When the distance is halved:

$$F_2 = \frac{GMm}{\left(\frac{r}{2}\right)^2} = \frac{GMm}{\frac{r^2}{4}} = \frac{4GMm}{r^2}$$

But $F_1 = \dfrac{GMm}{r^2}$ so $F_2 = 4\,F_1$.

Sometimes, when solving real-world problems in physics, you will not have all the necessary information that you need. It is helpful to be able to make a reasonable estimate for the answer to the problem. This can be done using order-of-magnitude figures for the variables in the problem. The physicist Enrico Fermi was able to calculate the approximate strength of the first nuclear bomb detonated in 1945. Watching the fireball of the explosion from 16 km away, Fermi dropped some small pieces of paper. He watched how they fell before, during and after the explosion. He estimated that the strength of the bomb was approximately 10 kilotons. The actual value was 20 kilotons.

Knowledge check 3

The light from a laser has a wavelength of 643 nm and a diffraction grating has a distance between slits of 3.3 μm. Express these values in m using standard form.

Exam tip

Always check whether the values you are given in an exam question have a prefix. Convert these into standard form before you attempt to answer the question.

Exam tip

Multiple-choice questions will often require you to make use of ratios to decide how changing one of the variables in an equation will affect another variable.

Knowledge check 4

The refractive index of olive oil is 1.47 and the refractive index of water is 1.33. Calculate the ratio speed of light in water (c_w) : speed of light in olive oil (c_o).

Worked example

A drop of vegetable oil, 0.5 mm in diameter, is dropped onto the surface of a tray of water. The oil spreads out across the water in a circular shaped layer with a diameter of 250 mm. Assuming that the layer is one molecule thick, estimate the size of a vegetable oil molecule.

Answer

The volume of the drop of vegetable oil remains the same whether it is as a drop or as the circular layer.

Drop volume — approximate the drop to a cube:

$$\text{volume} = (0.5 \times 10^{-3}\,\text{m})^3 \approx 1 \times 10^{-10}\,\text{m}^3$$

Layer volume — approximate the layer of oil on water to a square:

$$\text{volume} = (0.25\,\text{m})^2 \times \text{thickness} \approx (1 \times 10^{-1})\,\text{m}^2 \times d$$

where d is the thickness of one molecule.

Equating the two volumes gives:

$$1 \times 10^{-10}\,\text{m}^3 = 1 \times 10^{-1} \times d$$

So $d \approx 1 \times 10^{-9}\,\text{m}$.

This is an estimate of the largest size that a molecule may be, because the film may be more than one molecule thick.

At A-level, exponentials and logarithms are mainly found in the topics of capacitance and radioactivity. Exponential growth occurs when the variable you are measuring increases by the same proportion in each equal interval of time. Exponential decay occurs when the measured variable decreases by the same proportion in each equal interval of time.

Worked example

(A-level only) The potential difference (p.d.) V (in V) across a capacitor of capacitance C (in F) that is discharging through a resistor of resistance R (in Ω) decreases exponentially with time t (in s) according to the equation:

$$V = V_0 e^{-t/RC}$$

where V_0 is the initial p.d. at t_0.

A student is using data logging equipment to investigate the discharge of a capacitor. The capacitor has a labelled capacitance of 220 µF and is charged to 8.0 V. The capacitor is discharged through a 470 kΩ resistor. Calculate the p.d. across the capacitor after 60 s.

→

a If X is 120, what is 40% of X?
b If 20% of X is 60, what is X?
c If 25% of X is 2.0×10^3, what is X?

A student is determining the refractive index of a transparent block. Using a protractor, the student measures the angle of incidence as 35.5° and the angle of refraction as 26.0°. Calculate the refractive index of the block.

What are the values of $\sin\theta$, $\cos\theta$ and $\tan\theta$ for the following angles?
a $\theta = 0°$; b $\theta = 45°$;
c $\theta = \pi\,\text{rad}$; d $\theta = \pi/2\,\text{rad}$

If a question involves angles in radians, make sure that you use your calculator in the radian or rad mode for your calculations.

Answer

Using:

$V = V_0 e^{-t/RC}$ where $V_0 = 8.0\,\text{V}$ gives:

$$V = 8.0\,\text{V} \times e^{-\left(\frac{60\,\text{s}}{470\times10^3\,\Omega\,\times\,220\times10^{-6}\,\text{F}}\right)} = 8.0\,\text{V} \times e^{-0.580}$$

$$V = 8.0\,\text{V} \times 0.560 = 4.5\,\text{V}$$

Handling data

During most practical work you will collect data of some form. You will need to process the data in a suitable way to allow you to draw conclusions about the experiment.

Throughout your studies you will need to:

- use an appropriate number of significant figures (SF)
- find arithmetic means
- understand simple probability
- make order-of-magnitude calculations
- identify uncertainties in measurements and use simple techniques to determine uncertainty when data are combined by addition, subtraction, multiplication, division and raising to powers

Significant figures are the figures in a number that are meaningful or useful. When doing a practical, the number of significant figures recorded in your data will depend on the precision of the equipment you are using. You can find more detail about precision and uncertainties on pages 21–24.

Using experimental data to calculate other variables with your calculator will often result in an answer with a lot of decimal places. When writing out the answers, round up or down to give the answer to the same number of significant figures as the original data.

Worked example

A student is measuring the Young modulus of a copper wire. She takes the following measurements of the diameter of the wire at five different places along the wire using a micrometer:

0.37 mm	0.38 mm	0.36 mm	0.38 mm	0.37 mm

Calculate:

a the mean diameter of the wire

b the cross-sectional area of the wire

Give your answers to an appropriate number of significant figures.

→

Exam tip

Always check your answer to see if it is sensible and of the right magnitude. For example, if the value for the voltage after 60 s is larger than the initial value of 8.0 V, you have probably not used the minus sign in the exponential.

Knowledge check 8

Do the following calculations, giving your answers in standard form to the appropriate number of significant figures.

a $\frac{4}{3}\pi\,(6.37\times10^6)^3$

b $430\,\text{THz} \times 700\,\text{nm}$

c $(0.50 \times 6.0) + (\tfrac{1}{2} \times 9.8 \times (6.0)^2)$

Exam tip

How many significant figures? When carrying out a calculation using measured data with different numbers of significant figures, your result should contain the same number of significant figures as the measurement with the *smallest* number of significant figures (unless a detailed analysis of the uncertainties suggests otherwise).

Answer

a   mean diameter = $\dfrac{(0.37+0.38+0.36+0.38+0.37)\,\text{mm}}{5}$

   mean diameter = $0.372\,\text{mm} = 3.72 \times 10^{-4}\,\text{m}$

   However, while our calculator gives us 3 SF, our data have only 2 SF, so the answer should be given as:
   mean diameter = $3.7 \times 10^{-4}\,\text{m}$

b   cross-sectional area = $\pi r^2 = \pi(½ \times 3.72 \times 10^{-4}\,\text{m})^2 = 1.086\,87 \times 10^{-7}\,\text{m}^2$
   Again, rounding down to 2 SF, the answer should be:
   cross-sectional area = $1.1 \times 10^{-7}\,\text{m}^2$

Worked example

A student measures the potential difference across a resistor that is being used as part of a potential divider circuit. The potential difference is measured as 8.0 V. The current in the resistor is 2.96 mA. What is the resistance of the resistor?

Answer

$$R = \frac{V}{I} = \frac{8.0\,\text{V}}{2.96 \times 10^{-3}\,\text{A}} = 2702.7\,\Omega$$

The data in the question are given to 2 SF and 3 SF. The answer should therefore be given to just 2 SF. However, that would be $2700\,\Omega$, which could be taken to mean 2, 3 or 4 SF. In this case, it is more useful to give the resistance as $2.7\,\text{k}\Omega$.

Probability is a measure of the likelihood of an event occurring. In physics, you will mainly meet probability in the topic of radioactive decay. The probability of a nucleus decaying in a unit time is constant, known as the decay constant, λ. This gives us the equation for the *rate* of decay (or activity A) as $\dfrac{dN}{dt} = -\lambda N$, the minus sign indicating that the activity *decreases* with time.

Algebra

Algebraic equations are used to clearly express the relationship between variables in a physical situation. To solve many problems given in AS and A-level physics, you will need to be able to manipulate equations.

Throughout your studies you will need to:
- understand and use the symbols =, <, ≪, ≫, >, ∝, ≈, Δ
- change the subject of an equation, including non-linear equations
- substitute numerical values into algebraic equations using appropriate units for physical quantities
- solve algebraic equations, including quadratic equations
- **use logarithms in relation to quantities that range over several orders of magnitude**

Content Guidance

Symbols are a useful form of shorthand in physics. You will be familiar with many of these symbols from earlier in your school career. Some are shown in Table 3.

Table 3 Some symbols you should know

Symbol	Meaning
<<	much less than
>>	much greater than
∝	proportional to
≈	approximately equal to
Δ	change in

Exam tip

You are less likely to make a mistake in calculations if you rearrange the appropriate equations *before* you put the numbers in.

Worked example

(**A-level only**) A student is measuring the specific heat capacity of a block of aluminium. She uses a small heater to heat the block and a thermometer inserted into the block to measure the temperature rise. She obtains the following values:

Mass of block/kg	Energy supplied/J	Initial temperature/°C	Final temperature/°C
0.996	2.6×10^4	18	48

Calculate the value of specific heat capacity for the block of aluminium.

Answer

energy supplied = mass × specific heat capacity × temperature change

$$E = mc\Delta\theta$$

First divide both sides by $m\Delta\theta$ to get:

$$\frac{E}{m\Delta\theta} = \frac{mc\Delta\theta}{m\Delta\theta}$$

Cancelling $m\Delta\theta$ from the right-hand side of the equation we get:

$$\frac{E}{m\Delta\theta} = c$$

We can now substitute the values from the experiment into the equation:

$$c = \frac{2.6 \times 10^4 \text{J}}{0.966 \text{kg} \times 30 \text{K}} = 897.17 \text{Jkg}^{-1}\text{K}^{-1}$$

The smallest number of significant figures in the data recorded by the student is two. The answer should therefore be given to 2 SF:

$$c = 9.0 \times 10^2 \text{Jkg}^{-1}\text{K}^{-1}$$

Knowledge check 11

Write the following statements as equations using the appropriate mathematical symbols:

a The work done in an electrical circuit is equal to the potential difference multiplied by the charge.

b The extension of a spring is directly proportional to the applied force on the spring.

c **Magnetic flux is equal to the magnetic flux density multiplied by the area normal to the magnetic field.**

Exam tip

Remember that a temperature *difference* in °C is the same as a temperature *difference* in K. For example, here we have $(48 - 18)°C \equiv 30 \text{K}$.

Knowledge check 12

Rearrange the following equations so that the quantity shown in square brackets after each equation is the subject of the rearranged equation.

a $\rho = \frac{m}{V}$ [V]

b $p = mv$ [m]

c $V_{out} = V_{in} \times \frac{R_2}{R_1 + R_2}$ [R_2]

d $v^2 = u^2 + 2as$ [a]

e $n\lambda = d\sin\theta$ [θ]

f $F = m\omega^2 r$ [θ]

10 Edexcel Physics

Knowledge check 13

a Use the equation:

moment = Fd

to calculate the moment when F = 225 N and d = 7 cm.

b Use the equation:

$\varepsilon = \dfrac{x}{L}$

to calculate L when ε = 0.1 and x = 10 cm.

c **Use the equation:**

$x = A\cos\omega t$

to calculate x when A = 0.040 m, $\omega = \pi/3\,\text{rad s}^{-1}$ and t = 11 s.

Quadratic equations contain squared terms. The most common quadratic equation that you will meet in AS or A-level physics is the equation of motion:

$$s = ut + \frac{1}{2}at^2$$

Logarithmic scales are used to express data that have a range of several orders of magnitude. Sound is one common phenomenon that is measured using a logarithmic scale. Sound intensity is measured in W m^{-2}. A typical person can hear very quiet (low-intensity) sounds and extremely loud (high-intensity) sounds. To make a noise *sound* twice as loud, the intensity has to be approximately 10× greater. It would be very difficult to use a linear scale to represent this. A special logarithmic scale, known as the decibel (dB) scale, is used (Table 4). On this scale, a doubling of intensity is given by an increase of 10 dB.

Table 4 Sound intensity on dB scale

Sound	Intensity/W m^{-2}	Intensity/dB
Whisper	1×10^{-10}	20
Bird call	1×10^{-8}	40
Normal speech	1×10^{-6}	60
Traffic noise	1×10^{-5}	70
Pop/rock concert	1×10^{-1}	110
Pain threshold	1×10^{1}	130

Quantity algebra

Throughout this student guide use has been made of **quantity algebra**. This means giving *every* quantity in an equation its appropriate unit, e.g.:

$$\rho = \frac{6.95\,\Omega \times 7.79 \times 10^{-8}\,\text{m}^2}{0.500\,\text{m}} = 1.08 \times 10^{-6}\,\Omega\,\text{m}$$

Using quantity algebra gives a useful check for units and can often prevent silly mistakes.

Graphs

It has been said that a picture is worth a thousand words. This is definitely true in physics, where graphs are used to display data, to investigate relationships between variables and to calculate the magnitude of variables.

Throughout your studies you will need to:

- translate information between graphical, numerical and algebraic forms
- plot two variables from experimental or other data

Knowledge check 14

A ball is dropped from a height of 1.5 m. How long does it take the ball to reach the ground?

Knowledge check 15

(A-level only) Suggest another physical phenomenon that is measured using a logarithmic scale.

Exam tip

Get into the habit of using quantity algebra so that you can check whether your units are correct.

Content Guidance

- understand that $y = mx + c$ represents a linear relationship
- determine the slope and intercept of a linear graph
- calculate rate of change from a graph showing a linear relationship
- draw and use the slope of a tangent to a curve as a measure of rate of change
- distinguish between instantaneous rate of change and average rate of change
- understand the possible physical significance of the area between a curve and the x-axis and be able to calculate it or estimate it by graphical methods as appropriate
- apply the concepts underlying calculus to solve equations involving rates of change using a graphical method or spreadsheet modelling
- **interpret logarithmic plots**
- **use logarithmic plots to test exponential and power-law variations**
- sketch physical relationships that are modelled by:
  - $y = k/x$
  - $y = kx^2$
  - $y = kx$
  - $y = \sin x$
  - $y = \cos x$
  - $y = e^{\pm x}$
  - $y = \sin^2 x$
  - $y = \cos^2 x$

In general, when plotting graphs, the independent variable is plotted on the x-axis, and the dependent variable is plotted on the y-axis. You can read more about the practical skills involved in using graphs on pages 40–48.

A straight-line graph has the form $y = mx + c$, where m is the gradient of the graph and c is a constant. In practical work, the aim is to rearrange the equation describing the relationship between the variables into this form. If the relationship is correct, then your plot will be a straight line. Plotting data in this form also allows anomalous points to be seen more easily.

Worked example

A student is using a model of a rollercoaster track (Figure 1) to investigate the relationship between the initial height of a ball on the track and its speed at the top of the loop.

Release point

Light gate

Change in height, Δh

Data logger

Figure 1

> **Exam tip**
>
> Current–voltage characteristics and Hooke's law graphs are often plotted with the dependent variable on the x-axis. This is convenient as it makes the gradient equal to the constant of proportionality in each case (resistance and spring constant, respectively). Look carefully at the axes on graphs in case this has been done.

The student varies the initial height of the ball and measures the speed at the top of the loop using a light gate and data logging software. The initial energy of the ball is equal to its gravitational potential energy $E_p = mg\Delta h$. At the top of the loop its kinetic energy is $E_k = \frac{1}{2}mv^2$.

a What quantities should the student plot in order to obtain a straight-line graph?

b What does the gradient of the graph represent?

Answer

a In this case, $\Delta E_k = \Delta E_p$ so she can write:

$$mg\Delta h = \frac{1}{2}mv^2$$

Rearranging to get an equation in the form $y = mx + c$:

$$v^2 = 2g\Delta h$$

The student will need to plot v^2 (on the y-axis) against Δh (on the x-axis) to obtain a straight-line graph.

b The gradient is $m = 2g$, where g is the acceleration of free fall.

The gradient of any graph represents the change in y for a corresponding change in x:

$$\text{gradient} = \frac{\Delta y}{\Delta x}$$

This is a *rate of change*, and forms the basis of calculus.

To calculate the gradient from a straight-line graph, we draw a large triangle using the line of best fit as the hypotenuse of the triangle, as shown in Figure 2. In this case the gradient is equal to the rate of change of velocity with time $\left(\dfrac{\Delta v}{\Delta t}\right)$, or acceleration.

Figure 2

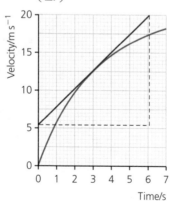

Figure 3

To calculate the gradient at a particular point from a curved graph, draw a tangent to the curve at that point. Then draw a large triangle using the tangent as the hypotenuse, as shown in Figure 3, and use this to calculate the gradient.

Mathematically, the gradient at a point is given by $\dfrac{dy}{dx}$ and can be found by differentiation.

Knowledge check 16

For the following relationships, state the quantities that need to be plotted to obtain a straight-line graph. In each case, state what is represented by the gradient of the graph.

a $c = f\lambda$ for variables f and λ

b $R = \dfrac{\rho l}{A}$ for variables R and l

c $T = 2\pi\sqrt{\dfrac{m}{k}}$ for variables T and m

Content Guidance

Figures 2 and 3 both show the motion of an object that is accelerating, but in Figure 3 the acceleration is changing. The gradient at a point represents the instantaneous acceleration of the object at that point. To calculate the average acceleration of the object, the readings over the whole of the graph need to be used.

As well as the gradient of a graph enabling us to determine a rate of change, the area under a graph may also have physical significance. In velocity–time graphs, such as Figures 2 and 3, the area under the graph represents the distance travelled. In order to calculate the appropriate area under the graph, your plot must include the origin.

Exam tip

When taking data from graphs, always read the units from the axes. Take care if they include prefixes, and make sure you use the correct values during your calculations.

(A-level only) Sometimes, it is helpful to plot data using a logarithmic plot because the data cover a wide range of values. One way to do this is to use special log graph paper. Figure 5 shows the radioactive decay of strontium with time, plotted on log-linear paper. Although the decay is exponential, the graph is linear because the mass has been plotted on a logarithmic scale.

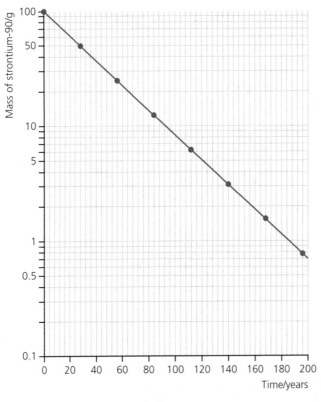

Figure 5

Instead of using special graph paper, a logarithmic plot of an exponential can also be drawn by taking natural logarithms ('ln' on your calculator) of both sides of the equation that describes the physical phenomenon.

Knowledge check 17

Using Figure 3 calculate the instantaneous acceleration of the object at 3 s, and the average acceleration of the object between 0 s and 6 s.

Knowledge check 18

State the physical quantity given by the area under each graph in Figure 4 and estimate the value of the quantity given by this area.

Figure 4

Worked example

(A-level only) A student measures the current, I, in a circuit as a capacitor discharges over time, t.

a What does she have to plot to obtain a straight-line graph?

b What do the intercept and gradient of the graph represent?

Answer

a The equation that describes the current is:

$$I = I_0 e^{-t/RC}$$

Taking natural logarithms ('ln') of both sides gives:

$$\ln I = \ln I_0 - \frac{t}{RC}$$

Plotting $\ln I$ against t will give a straight-line graph, as R and C are constants.

b The gradient is equal to $-\dfrac{1}{RC}$.

The intercept is equal to $\ln I_0$, where I_0 is the current at $t = 0$.

Figure 6 gives examples of the shapes of graphs for a range of different functions.

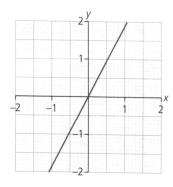

(a) Graph of $y = kx$, where k is a constant

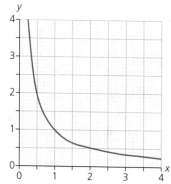

(b) Graph of $y = \dfrac{k}{x}$, where k is a constant

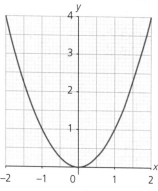

(c) Graph of $y = kx^2$, where k is a constant, in this case $k = 1$

(d) Graph of $y = \dfrac{k}{x^2}$, where k is a constant, in this case $k = 1$

> **Exam tip**
>
> The graph of any exponential function of the form $y = a e^{bx}$ will give a straight line when we plot $\ln y$ against x.
>
> The graph of a power function $y = ax^b$ will give a straight line when we plot a graph of $\ln y$ against $\ln x$, and the gradient of the line will be equal to b.

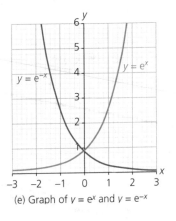

(e) Graph of $y = e^x$ and $y = e^{-x}$

(f) Graph of $y = \sin x$ and $y = \cos x$

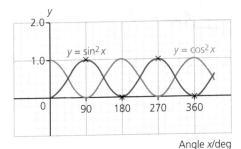

(g) Graph of $y = \sin^2 x$ and $y = \cos^2 x$

Figure 6

Knowledge check 19

For each of the following relationships, state which sketch graph from Figure 6 they would produce.

a Q against V for $Q = CV$

b E_k against v for $E_k = \frac{1}{2}mv^2$

c g against r for $g = \dfrac{GM}{r^2}$

Geometry and trigonometry

Modelling the real world in physics requires us to solve problems in two (2D) or three (3D) dimensions. For example, we may need to think about forces acting at angles to one another; or the surface area of a solar panel normal to the sunlight.

Throughout your studies you will need to:

- use angles in regular 2D and 3D structures
- visualise and represent 2D and 3D forms, including 2D representations of 3D objects
- calculate areas of triangles, circumferences and areas of circles, and surface areas and volumes of rectangular blocks, cylinders and spheres
- use Pythagoras' theorem, and the angle sum of a triangle
- use sin, cos and tan in physical problems
- **use small-angle approximations, including:**
  - **$\sin\theta \approx \theta$**
  - **$\tan\theta \approx \theta$**
  - **$\cos\theta \approx 1$ for small values of θ**
- understand the relationship between degrees and radians, and translate from one to the other

When doing practical work, or analysing data from practical work, you will often have to calculate different geometric properties of objects. Table 5 gives some useful formulae for this.

Table 5 Some useful formulae for geometric properties

Calculation	Shape	Equation
Area under an F–x graph for a spring obeying Hooke's law	Triangle	area = ½ × base × height
Cross-sectional area of a wire	Circle of radius r	area = πr^2
Energy radiated by an incandescent bulb	Sphere of radius r	surface area = $4\pi r^2$
Energy radiated by a fluorescent tube	Cylinder of radius r and height h	surface area = $2\pi r^2 + 2\pi rh$
Density of a block	Rectangular block of sides a, b and c	volume = $a \times b \times c$
Density of a marble	Sphere of radius r	volume = $\frac{4}{3}\pi r^3$
Volume of a gas cylinder	Cylinder of radius r and height h	volume = $\pi r^2 h$

When calculating the resultant of forces that are applied to an object, knowledge of trigonometry is essential.

Pythagoras' theorem states that, if a right-angled triangle has sides a, b and c, where c is the hypotenuse, then $a^2 + b^2 = c^2$ (Figure 7). Right-angled triangles are also used to define the three functions sine, cosine and tangent.

- $\sin\theta$ = opposite/hypotenuse
- $\cos\theta$ = adjacent/hypotenuse
- $\tan\theta$ = opposite/adjacent

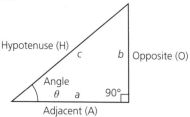

Figure 7

Worked example

A student is investigating equilibrium using a force board as shown in Figure 8.

Figure 8

The system comes to rest with angles $\theta_1 = 57°$ and $\theta_2 = 42°$. Show that the system is in equilibrium.

> **Exam tip**
>
> Data are often given for the diameter of a circle or a sphere, instead of the radius. Remember that radius = ½ × diameter and so $r^2 = \frac{1}{4}d^2$.

> **Knowledge check 20**
>
> A copper pipe has an internal diameter of 12 mm and a length of 150 cm. Calculate the volume of liquid that can be contained in the pipe.

> **Knowledge check 21**
>
> The molar mass of carbon is $12.0 \times 10^{-3}\,kg\,mol^{-1}$ and the density of diamond is $3500\,kg\,m^{-3}$. Estimate the atomic radius of a spherical carbon atom.

> **Exam tip**
>
> The sum of the internal angles in a triangle is 180°.

Answer

To be in equilibrium, the forces in each direction at point O (weight 3) must balance.

Resolving horizontally: $T_1 \sin \theta_1 = T_2 \sin \theta_2$

$T_1 \sin \theta_1 = 4\,\text{N} \sin 57° = 3.4\,\text{N}$ and $T_2 \sin \theta_2 = 5\,\text{N} \sin 42° = 3.3\,\text{N}$

We can see that the forces do not balance exactly. This is probably due to friction in the system causing experimental uncertainty.

Resolving vertically: $T_1 \cos \theta_1 + T_2 \cos \theta_2 = W_3$

$4\,\text{N} \cos 57° + 5\,\text{N} \cos 42° = 5.9\,\text{N}$

Once again, the vertical components balance to within a reasonable experimental error.

Degrees and radians are used to describe angles. In many cases, the radian is a useful measure, particularly when working with waves and oscillations.

A radian is defined as the angle created by the arc s of a circle that is equal in length to the radius r of the circle ($s = r$). Figure 9 shows that an angle θ in radians is given by $\theta = s/r$ or $s = r\theta$.

When using small angles, measured in radians, we can use the following **small-angle approximations**, as the values are very similar:

$\sin \theta \approx \theta$

$\tan \theta \approx \theta$

$\cos \theta \approx 1$

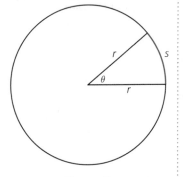

Figure 9

Exam tip

1 radian = 57.3°

Angle in radians =
$\dfrac{\pi}{180°}$ × angle in degrees

Angle in degrees =
$\dfrac{180°}{\pi}$ × angle in radians

Knowledge check 22

Express the following angles in radians to 3 SF: **a** 15.0°; **b** 50.0°; **c** 145°.

Knowledge check 23

Express the following angles in degrees to 3 SF: **a** $\pi/8$ rad; **b** 1.50 rad; **c** 0.150 rad.

Knowledge check 24

Calculate $\sin \theta$, $\cos \theta$ and $\tan \theta$ to 2 SF for the following angles: **a** 5.0 rad; **b** 0.50 rad; **c** 0.050 rad.

Exam tip

Make sure you have your calculator set in the right mode — degrees or radians — as required by the question.

Practical skills

Physics is a practical subject, and experimental work should form a significant part of your AS or A-level physics course. It is commonly acknowledged that it is easier to learn and remember things if you have actually done them rather than having read about them or been told about them. That is why this section of the book is illustrated throughout by experiments, usually with a set of data for you to work through and questions to answer.

As you complete practical work during your physics course you will develop skills in the use of a range of apparatus and techniques. Examination questions will test your knowledge and understanding of these techniques. You should refer to the specification, which contains a list of apparatus and techniques with which you should be familiar.

About this section of the book

For convenience, this section of the practical guide is in five main sections, which reflect an orderly approach to practical work:

■ Planning
■ Making measurements
■ Recording data
■ Analysing results
■ Evaluating results and drawing conclusions

Having said this, there will inevitably be a degree of overlap between each section.

Planning

Once the problem has been identified, you will be expected to produce a plan. Planning an experiment or investigation includes selecting appropriate equipment to make the relevant measurements. You will also need to be able to describe measurement strategies and techniques to ensure accurate results. Safety issues must also be discussed.

Types of variables

You are expected to be able to identify variables, including those that must be controlled. You should be familiar with the following terminology:

■ The *independent variable* is the variable for which values are changed by the experimenter.
■ The *dependent variable* is the variable for which the value is measured for each and every change in the independent variable.
■ *Control variables* may, in addition to the independent variable, affect the outcome of the investigation and therefore have to be kept constant or at least monitored.

Worked example

A student investigates how the diameter of a ball bearing affects how long it takes to fall through a measuring cylinder filled with oil. What are the independent, dependent and control variables in this experiment?

Answer

Independent variable is diameter of ball bearing.

Dependent variable is time.

Control variables are temperature of oil, distance fallen, material of ball bearing, smoothness of ball bearing surface, type of oil and height of release of ball bearing.

A key part of planning an experiment is describing how to obtain accurate and precise data. This involves choosing the most appropriate equipment and using it in such a way to enable the most accurate data to be obtained. We will look at good experimental design on page 26 after the section on errors and uncertainties.

Making measurements

Errors, accuracy and precision

One of the specified requirements is that you should *consider margins of error, accuracy and precision of data.* In everyday English, accuracy and precision have similar meanings, but in physics this is not the case. Their meanings are not the same and the difference must be understood.

An error is the difference between a measured result and the true value or accepted value, i.e. the value that would have been obtained in an ideal measurement (which is impossible to achieve!). With the exception of a fundamental constant, the true value is considered unknowable. An error can be due to random or systematic effects, and an error of unknown size is a source of *uncertainty*.

Random errors

There are always random errors present due to the way any instrument works, the way it is used or changes in the external conditions. The same measurement will give different values each time it is measured. The effect of this type of error can be reduced by taking repeat measurements and then averaging or by drawing a graph. Random errors produce scatter of the points around the line of best fit (Figure 10).

Random errors will give rise to an uncertainty in the measurement you have taken. It is important to be able to estimate the uncertainty in a measurement so that its effect can be taken into consideration when drawing conclusions about experimental results.

Systematic errors

A systematic error produces measurements that are consistently too large or too small by the same amount. This could be caused by recording the wrong unit, poor technique (e.g. systematically reading off from a scale at an angle, causing parallax error), failure to check for zero error, or incorrect calibration of the instrument.

Zero error

A zero error occurs when the measuring instrument is not set on zero accurately. The zero error should be subtracted from or added to each measured reading to give the true value (Figure 11).

| (a) No zero error | (b) Positive zero error | (c) Negative zero error |

Figure 11

Sometimes a systematic error is not apparent from the measurements. Another advantage of drawing a graph is that it can indicate whether there is a systematic error and enable you to make allowance for it (Figure 12).

Figure 10

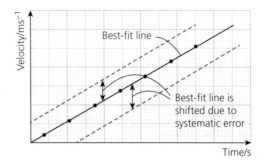

Figure 12

A systematic error will be seen on a graph by the line of best fit being shifted up or down the y-axis because each measurement is larger or smaller than the true value by the same amount. A systematic error does *not* affect the *gradient* of the line of best fit but simply alters the y-intercept.

Accuracy and precision

A measured value is considered to be **accurate** if it is judged to be close to the true value. Any actual measurement will always be subject to random and systematic errors.

The term **precision** denotes the consistency between values obtained by repeated measurements — a measurement is precise if the values 'cluster' together. Precision is influenced only by random errors. We can try to ascertain and quantify these errors, i.e. *estimate the uncertainty* in our value. Precise results would show little scatter around the line of best fit on a graph but might have a systematic error making them inaccurate. Accuracy and precision are summarised in Figure 13.

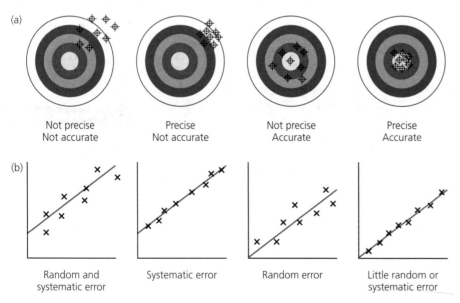

Figure 13

Knowledge check 27

Explain the difference between accuracy and precision.

Knowledge check 28

The magnetic field strength of the Earth at a point on the Earth's surface was measured five times and gave the following values: 50 mT, 48 mT, 46 mT, 52 mT, 62 mT. The true value of the magnetic field strength at this point was known to be 51 mT. Comment on the precision of these data.

Content Guidance

Uncertainties

Every measurement has an inherent uncertainty due to the resolution of the instrument, manufacturing error, the way the instrument is used (human error!) or the design of the experiment.

An uncertainty for a single reading, or repeated readings that are the same, will be at least half the **resolution** of the instrument. For repeated readings that are different, the uncertainty can be taken as being *half the range* of the readings.

As a general guide, the uncertainty of a **reading** is at least ±0.5 of the resolution. A reading is a value found from a single judgement, e.g. when reading a thermometer, Geiger counter, top-pan balance, measuring cylinder, or ruler with a zero end. For such 'analogue' devices, it may be possible to interpolate to the nearest half a division on the scale (Figure 14).

The uncertainty of a **measurement** is at least ±1 of the resolution. A measurement requires the value to be taken from two readings or judgements, e.g. when using a non-zero-ended ruler (the start and the end readings), stopwatch (start and stop) or extension of a spring (original and final lengths).

Figure 14

Exam tip

When using an analogue instrument, always try to interpolate between whole divisions, e.g. to the nearest 0.5 mm on a ruler or to 0.1 degrees for a thermometer calibrated in degrees.

Worked example

Estimate the uncertainty in a length measurement of 122 mm made using the ruler shown in Figure 15, and write down how the length should be recorded.

Figure 15

Answer

The smallest division on the ruler is 1 mm. Therefore, the uncertainty in each reading is ½ × 1 mm = 0.5 mm.

But as we are taking readings at each end, we are making two judgements. So the total uncertainty in this measurement of length would be ±(2 × 0.5 mm) = ±1 mm.

length = 122 mm ± 1 mm

Other factors affecting the uncertainty in a measurement

The general guidelines above give the *smallest* uncertainty in a measurement or reading. Often the resolution of the instrument is not the limiting factor, and the way the instrument is used can lead to larger uncertainties. The instrument's resolution allows us to assess the *minimum* possible uncertainty. You need to make an assessment of the other factors, such as the set-up and use of the apparatus, and take these into account.

In experiments, you will often be timing events or measuring the length of something. The following examples should help you understand how the uncertainties involved are actually much larger than those due to the resolution of the stopclock or ruler.

Exam tip

In exam papers, values will often be stated together with the absolute uncertainty, e.g. 'a wire has a diameter of 1.87 mm ± 0.01 mm'. The uncertainty should be quoted to be the same number of decimal places as the reading.

Knowledge check 29

What is meant by the resolution of an instrument?

Exam tip

Always write down the full reading and then, if necessary, round to fewer significant figures when the uncertainty has been estimated.

A stopwatch typically has a resolution of 0.01 s, but the reaction time of the person using it is likely to be no quicker than 0.05 s or 0.10 s. You should record the full reading on the stopwatch (e.g. 18.62 s) and reduce this to a more appropriate number of significant figures at a later stage, e.g. after averaging readings.

If you are measuring the length of a piece of an unstretched elastic band, it is difficult to place the elastic completely straight against the ruler. This could lead to an uncertainty of ±2 or ±3 mm.

The uncertainty of the reading from digital voltmeters and ammeters depends on the tolerance quoted by the manufacturer (which may be considerable). If this is not known, then the reading should usually be stated to ±1 in the last digit when quoting the value and the uncertainty.

Knowledge check 30

A 100 Ω resistor has a tolerance of 5%. Calculate the minimum and maximum values it could have.

Worked example

Estimate the uncertainty in the reading of potential difference shown in Figure 16, and state how you would record the value of the p.d. The dial is set on the 20 V dc scale.

Figure 16

Answer

The uncertainty will be ± the smallest division shown on the meter = ±0.01 V.

The reading is therefore 19.16 V ± 0.01 V.

The manufacturing tolerance is not stated, so by quoting the uncertainty to ± the smallest division, we are making a best possible estimate.

Percentage uncertainties

It is useful to be able to quote uncertainties as percentages. This also enables uncertainties to be combined easily.

$$\text{percentage uncertainty} = \frac{\text{uncertainty}}{\text{value}} \times 100\%$$

Worked example

The digital voltmeter shown in Figure 16 is being used to measure the p.d. across a lamp. Estimate the percentage uncertainty in the reading of 19.16 V.

Answer

The uncertainty will be ± the smallest division shown on the meter = ±0.01 V.

The reading is therefore 19.16 V ± 0.01 V.

Percentage uncertainty = (0.01 V/19.16 V) × 100% = 0.05%

Combining uncertainties

In experiments, measurements are often made using several instruments and then used to calculate a quantity. The combined uncertainty can be found as follows:

- If quantities are added or subtracted, *add* the *absolute* uncertainties of each quantity.
- If quantities are multiplied or divided, *add* the *percentage* uncertainties of each quantity.
- If a quantity is raised to a power, then *multiply* the *percentage* uncertainty by the *power*.

Worked example

The p.d. across a bulb is measured to be 3.15 V when the current in it is 40.8 mA.
a Calculate the power.
b Estimate the percentage uncertainty in the power.
c State how the power should be recorded.

Answer

a Power = VI = 3.15 V × 40.8 × 10^{-3} A = 0.129 W (data to 3 SF, and so answer should be to 3 SF)

b The p.d. and current are both *readings*, so the uncertainty is *theoretically* half the smallest division, but in practice we take it to be the smallest division to make some allowance for manufacturing tolerance of the digital meters. In this case, 0.01 V and 0.1 mA. (We shall often write % uc as shorthand for percentage uncertainty.)

So, % uc in V = (0.01 V/3.15 V) × 100% = 0.317%

and % uc in I = (0.1 mA/40.8 mA) × 100% = 0.245%

This gives % uc in P = % uc in V + % uc in I

so % uc in P = 0.317 + 0.245 = 0.562% = 0.6%

c Uncertainty in P = 0.562% of 0.129 W = 7.2 × 10^{-4} W ≈ 0.001 W

Power = 0.129 W ± 0.001 W or 129 mW ± 1 mW

Recording data

You will be expected to *present data in appropriate ways*. You may also need to *process the data using appropriate mathematical skills*.

Exam tip

The manufacturer's tolerance will almost certainly be much greater than 0.05%, and for many digital meters it can be as much as 5% or even 10%.

Knowledge check 31

The diameter of a ball bearing is measured using vernier callipers with 0.1 mm divisions. The vernier reading is 21.5 mm.
a State the uncertainty in this measurement, and write down how the value for the diameter should be recorded.
b Estimate the percentage uncertainty in this value.

Exam tip

Keep the data in your calculator and only round the final answer.

Tabulating data

Data must be clearly recorded in a table. Tables should have headings with the quantity and units separated by a forward slash (e.g. current/mA). You should also ensure that the number of significant figures, or decimal places, is appropriate, and these should be consistent down each column. The independent variable should normally be the first column of the table.

Table 6, containing the results from an experiment to measure electrical power, shows how data should be recorded.

Table 6

Potential difference V/V	Current I/mA
6.00	58.8
4.63	50.8
3.31	41.8
2.16	32.9

Worked example

(**A-level only**) In an experiment to investigate whether radioactive decay is exponential, a student decides to plot a graph of the natural logarithm of the activity A, which is measured in counts per second, against time t, which is measured in seconds. Draw a table showing the headings that the student should use in order to record these data.

Answer

Time t/s	Activity A/s^{-1}	ln $(A$/s$^{-1})$

Significant figures

When recording data, it is important to *write down* the data to the number of significant figures given by the resolution of the device being used. For digital instruments, you should write down all of the numbers shown on the device, e.g. times should be *recorded* to 0.01 s if using a digital stopwatch. Only in the subsequent processing should you round down to the appropriate number of significant figures. The following examples show how to correctly use significant figures in recorded and processed data.

Worked example

In an experiment to find the period T of a simple pendulum, the following times are recorded for 20 oscillations:

t_1/s	t_2/s	t_3/s	Mean t/s	T/s	T^2/s^2
14.72	14.65	14.67			

Complete the table by adding values of t, T and T^2.

Answer

Mean t = 14.68 s $T = t/20 = 0.734$ s $T^2 = 0.539$ s^2

Knowledge check 32

What is wrong with this table?

Volts	A
5.85 V	0.05
4.38 V	0.04
2.94 V	0.03
1.4 V	0.01

Exam tip

If you are doing A-level physics, you will be dealing with exponential changes and logarithmic quantities. As logarithmic quantities do not have any units, you should write a table heading in the following format: ln (quantity/unit).

Exam tip

Remember, your answer should contain the same number of significant figures as the measurement with the *smallest* number of significant figures (unless a detailed analysis of the uncertainties suggests otherwise).

Exam tip

Don't forget to divide by 20 to find the period.

Content Guidance

Worked example

The following data were recorded for a trolley accelerating down an inclined plane:

■ initial velocity of the trolley = $1.23\,\mathrm{m\,s^{-1}}$
■ final velocity of the trolley = $1.97\,\mathrm{m\,s^{-1}}$
■ time taken = $1.6\,\mathrm{s}$

Use these data to calculate the acceleration of the trolley, quoting your answer to the appropriate number of significant figures.

Answer

$$a = (v - u)/t = (1.97 - 1.23)\,\mathrm{m\,s^{-1}}/1.6\,\mathrm{s}$$

$$= 0.4625\,\mathrm{m\,s^{-2}} = 0.46\,\mathrm{m\,s^{-2}}$$

(Answer is to 2 SF because the time is only recorded to 2 SF even though the velocities are recorded to 3 SF.)

Good experimental design

When planning an experiment, you should attempt to design a method that leads to the smallest uncertainties in your measurements. Exam questions may ask you to look at a number of factors:

■ the resolution of the instrument used
■ the manufacturer's tolerance on instruments and components such as resistors and capacitors
■ how the experimenter uses the instruments (e.g. parallax, zero error)
■ the procedures adopted (e.g. repeated readings)

You may be required to evaluate procedures and explain how particular techniques could affect uncertainties in the measurements, or how the uncertainties could be reduced by using different apparatus or procedures.

Parallax error

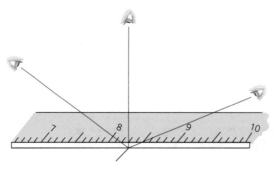

Figure 17

When reading a scale, for example from an analogue meter or ruler, you should ensure that the pointer, scale and your eye are all in line. This will avoid *parallax error*. It is important not to look at the scale from an angle, as this would give a measurement that is larger or smaller than the true value (Figure 17).

> **Knowledge check 33**
>
> In an experiment to measure the acceleration of a trolley, the acceleration was measured three times using two light gates and a data logger:
>
> | Acceleration/m s^{-2} | | |
> |---|---|---|
> | 0.44 | 0.45 | 0.47 |
>
> Calculate the mean acceleration, quoting your answer to the appropriate number of significant figures.

> **Knowledge check 34**
>
> a Figure 17 shows an enlarged millimetre scale. What is the correct reading?
> b Explain how a mirror can be used to reduce parallax error when taking a reading from an analogue scale that has a pointer.

Using a vernier scale

Vernier callipers and micrometers enable measurements of length or thickness to be made to a much greater resolution than using a ruler and can reduce the uncertainty by a factor of 10 or 100, respectively. Often you will be able to use electronic or digital instruments, which have an even greater resolution.

Vernier callipers measure to the nearest 0.1 mm and therefore measure to a precision of 1% or better if used to measure lengths of 1 cm or longer.

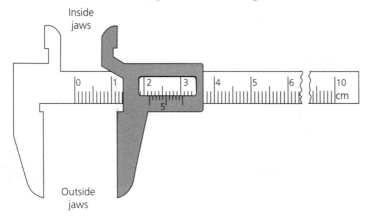

Figure 18

Figure 18 shows how to read the scale — a hand lens is recommended, as the scale lines can be difficult to see.

- the zero on the vernier scale is just beyond the 2.1 cm (21 mm) mark on the main scale
- next locate the lines on the main scale and the vernier scale that are *exactly lined up with each other*
- in this case this is the '5' on the vernier scale
- the reading is therefore 2.15 cm (21.5 mm)

Vernier scales are used in a number of instruments, for example travelling microscopes, but the principle of reading the scale is always the same.

The following example shows how, by using a vernier scale and a sensitive electronic balance, the accuracy of the measurements of length and mass, and therefore the determination of density, can be improved. This is particularly important when a quantity is being squared or cubed.

Worked example

a A student is asked to determine the density of aluminium in the form of a cylinder. He measures the diameter and length of the cylinder with a metre rule and the mass with a top pan balance. He records the following measurements:

   length = 75 ± 1 mm

   diameter = 25 ± 1 mm

   mass = 98.8 ± 0.1 g

   Use these data to determine the student's value for the density of aluminium, and estimate the percentage uncertainty in his value.

Knowledge check 35

What does the vernier scale shown in Figure 19 read?

Figure 19

Knowledge check 36

Calculate the density of a marble having a diameter of 10 mm and a mass of 1.3 g. Give your answer in $kg\,m^{-3}$ and in $g\,cm^{-3}$.

Exam tip

You are less likely to make a mistake if you let the calculator do the work when converting units rather than doing it in your head — put in the radius of 12.5 mm as $12.5 \times 10^{-3}\,m$ rather than trying to work out 0.0125 m.

Content Guidance

b His teacher suggests that he would get a more accurate value for the density if he used vernier callipers to measure the diameter and length.

What does the teacher mean by 'accurate'? Estimate the improvement to the accuracy this would make.

Answer

a volume = $\pi r^2 h = \pi \times (12.5 \times 10^{-3}\,\text{m})^2 \times 75 \times 10^{-3}\,\text{m} = 3.68 \times 10^{-5}\,\text{m}^3$

density = mass/volume = $98.8 \times 10^{-3}\,\text{kg}/3.68 \times 10^{-5}\,\text{m}^3$

$= 2.7 \times 10^3\,\text{kg m}^{-3}$ (2 SF in line with the data)

% uc in $d = (1\,\text{mm}/25\,\text{mm}) \times 100\% = 4.0\%$

so % uc in r^2 = % uc in $d^2 = 2 \times 4.0\% = 8.0\%$

% uc in $h = (1\,\text{mm}/75\,\text{mm}) \times 100\% = 1.3\%$

% uc in $m = (0.1\,\text{g}/98.8\,\text{g}) \times 100\% = 0.1\%$

% uc in density $= 8.0\% + 1.3\% + 0.1\% = 9.4\% \approx 9\%$

b An accurate value is one that is close to the true value.

A vernier scale will have a resolution of 0.1 mm compared with 1 mm for the rule. The percentage uncertainties in d and h will therefore be 0.40% and 0.13%, respectively.

The percentage uncertainty in the density will then be $(2 \times 0.40 + 0.13 + 0.10) = 1.03\% \approx 1\%$.

This means the accuracy will be improved by a factor of about nine times.`

Using a micrometer

A micrometer (Figure 20) has a resolution of ±0.01 mm. The micrometer should be closed using the ratchet until it *just* clicks.

Figure 20

When using a micrometer, remember the following:

- Do not over-tighten, as this could squash the object and/or damage the thread. Use the *ratchet* to close the jaws.
- Check for a zero error and add or subtract as necessary.
- Make sure you read the scale correctly (in particular, whether the main scale reading is less than or more than 0.5 mm).
- Measure the diameter several times and in several places to check for uniformity. (For a wire, for example, you could measure the diameter at each end and in the middle to check if the wire is thinner at one end. You could also then take readings at right angles to check that the wire has a circular cross-sectional area.)

Exam tip

When a quantity is raised to the power of n, the *percentage* uncertainty in the quantity is multiplied by n.

Knowledge check 37

a What is the resolution of the main scale of the micrometer shown in Figure 20?
b What is the thickness of the object being measured by the micrometer shown in Figure 20? What assumption are you making?

Exam tip

You should be able to state the precautions that should be taken to improve accuracy when using a micrometer.

Knowledge check 38

A micrometer has a zero error of +0.03 mm. A student measures the diameter of a pencil lead to be 2.12 mm. What is the actual diameter of the lead?

Repeated measurements

Random errors are generally caused either by human misjudgements, e.g. reaction time, interpolating between scale divisions, or by variations in a measurement, e.g. non-uniformity of a wire. Random errors will cause measurements to spread about a mean value. As precise measurements are measurements in which there is very little spread about the mean value, you must try to reduce random errors in order to improve precision.

Because random errors are likely to cause readings to be larger than the true value as often as smaller than the true value, repeating readings or measurements will smooth out the effect of random errors.

For repeat measurements:

$$\text{percentage uncertainty} = \frac{\frac{1}{2} \times \text{range}}{\text{mean value}} \times 100\%$$

By plotting a graph, the line of best fit will effectively average out the random errors, but they can never be eliminated completely. You will also be able to see the scatter of the points about the line of best fit and so see how precise the measurements are.

Worked example

The resistance of a thermistor was measured three times at 20°C: 670 Ω, 661 Ω, 678 Ω. Calculate the percentage uncertainty in the value of the resistance, and state how the resistance of the thermistor should be recorded.

Answer

range = 678 − 661 = 17 Ω

mean = (670 + 661 + 678)/3 = 669.7 Ω = 670 Ω (3 SF)

$\%\text{uc} = \dfrac{(\frac{1}{2} \times 17)\,\Omega}{670\,\Omega} \times 100\% = 1.27\%$

1.27% of 669.7 Ω = 8.5 Ω ⇒ R = 670 Ω ± 9 Ω

Worked example

For an angle of incidence of 40.0° on a glass block, the angle of refraction was measured three times as: 25.5°, 25.0°, 26.0°. How should the mean value of the angle of refraction be expressed?

Answer

range = 26.0° − 25.0° = 1.0°

mean = 76.5/3 = 25.5°

angle of refraction = 25.5° ± 0.5°

Improving precision: you need to be able to describe how the measurements obtained from an experiment can be made more precise.

Explain how the precision of experimental measurements can be improved.

Worked example

Measuring the density of glass and sand

We can illustrate a number of the points we have made concerning good experimental design by considering an experiment to investigate whether glass might be manufactured from sand. The following apparatus is used: glass block, plastic cup, dry sand, 250 ml measuring cylinder, vernier callipers and access to balance and water.

The experiment is carried out as follows:

For the glass block:

- Using the vernier callipers, the dimensions are determined as:

    $l = (116.0 \pm 0.2)\,\text{mm}$

    $w = (63.0 \pm 0.2\ )\,\text{mm}$

    $h = (48.7 \pm 0.2)\,\text{mm}$

- Using the electronic balance:

    $\text{mass} = 882.4 \pm 0.1\,\text{g}$

For the sand:

- Using the electronic balance:

    mass of empty plastic cup $= 35.5 \pm 0.1\,\text{g}$

    mass of cup and sand $= 87.1 \pm 0.1\,\text{g}$

- The measuring cylinder is filled with $100\,\text{cm}^3$ of water and then the sand is added.

    volume of the water + sand $= 123\,\text{cm}^3$

a Determine the densities of glass and sand.
b Estimate the uncertainties in each of these values.
c Make a comparison between your final values with reference to these uncertainties.
d Write a conclusion discussing the extent to which your results support the fact that glass is made from sand.
e Explain any experimental techniques that you would use in the above experiment to try to reduce the uncertainty in the measurements.

Answer

a *Glass:*

    $\text{volume} = 116.0 \times 10^{-3}\,\text{m} \times 63.0 \times 10^{-3}\,\text{m} \times 48.7 \times 10^{-3}\,\text{m}$

    $= 3.56 \times 10^{-4}\,\text{m}^3$

    $\text{density} = \text{mass/volume} = 882.4 \times 10^{-3}\,\text{kg}/3.56 \times 10^{-4}\,\text{m}^3$

    $= 2.48 \times 10^3\,\text{kg}\,\text{m}^{-3} \text{ or } 2.48\,\text{g}\,\text{cm}^{-3}$

→

Sand:

volume of sand $= (123 - 100)\,cm^3 = 23\,cm^3 \pm 2\,cm^3$ (as we have two readings)

mass of sand $= (87.1 - 35.5)\,g = 51.6\,g \pm 0.2\,g$

density = mass/volume $= 51.6\,g/23\,cm^3$

$\qquad = 2.2\,g\,cm^{-3}$ (to 2 SF, as volume is only to 2 SF)

b *Glass:*

% uc in mass $= (0.1\,g/882.4\,g) \times 100\% = 0.01\%$

% uc in volume = % uc in length + % uc in width + % uc in depth

$\qquad\qquad = (0.17 + 0.32 + 0.41)\% = 0.90\%$

% uc in density = % uc in mass + % uc in volume

$\qquad\qquad = (0.01 + 0.90)\% = 0.91\%$

Uncertainty in density = 0.91% of $2.48\,g\,cm^{-3} = \pm 0.02\,g\,cm^{-3}$

Sand:

% uc in density = % uc in mass + % uc in volume

$= (0.2\,g/51.6\,g) \times 100\% + (2\,cm^3/23\,cm^3) \times 100\%$

$= (0.4 + 8.7)\% = 9.1\%$

Uncertainty in density = 9.1% of $2.2\,g\,cm^{-3} = \pm 0.2\,g\,cm^{-3}$

c The percentage difference between the density of glass and the density of sand is:

$$\frac{(2.48 - 2.2)\,g\,cm^{-3}}{2.34\,g\,cm^{-3}} \times 100\% = 12\%$$

d The minimum value that the glass could have is $2.46\,g\,cm^{-3}$ and the maximum possible value for the sand is $2.4\,g\,cm^{-3}$.

As the values do not quite overlap, the 12% difference cannot be accounted for by the calculated experimental uncertainties, suggesting that glass may *not* be made entirely from sand. More likely reasons for the discrepancy are that the sand may not have been completely dry initially, or not all the air between the sand particles was replaced by water in the measuring cylinder.

e You might suggest the following:

■ Check vernier for zero error.

■ Repeat measurements of block dimensions in at least two different places for each dimension.

■ Shake the sand well in the measuring cylinder to prevent air pockets.

■ Keep eye level with water meniscus to prevent parallax error.

Knowledge check 40

Explain how you would determine, as accurately as possible, the volume of a glass marble.

Exam tip

The percentage difference between two experimental values is given by:

$$\%\text{difference} = \frac{\text{difference between the values}}{\text{average of the two values}} \times 100\%$$

The percentage difference between an experimental value and a stated or known value is given by:

$$\%\text{difference} = \frac{\text{difference between the values}}{\text{stated value}} \times 100\%$$

Multiple readings

The use of multiple readings reduces the uncertainty. For example, if you determine the thickness of one sheet of paper by measuring the thickness of 50 sheets in a stack, and dividing by 50, you will then reduce the uncertainty in the measurement of one sheet by a factor of 50. In general, the uncertainty of each measurement will be the uncertainty of the whole measurement divided by the number of instances (e.g. sheets of paper or oscillations of a pendulum). This method works because the *percentage* uncertainty in the value of the thickness of one sheet will be the same as the percentage uncertainty in the measurement of multiple sheets.

Worked example

The thickness of ten sheets of paper is measured with a micrometer (resolution ±0.01 mm) and found to be 1.08 mm. What is the percentage uncertainty in the mean thickness of one sheet?

Compare this to a measurement of the thickness of a ream of 500 sheets made using a vernier (resolution ±0.1 mm), which is 55.1 mm.

Answer

Thickness of ten sheets of paper: 1.08 mm ± 0.01 mm = 1.08 mm ± 0.9%

Mean thickness of one sheet of paper = 1.08 mm/10 = 0.108 mm ± 0.9%

Thickness of ream of 500 sheets = 55.1 mm ± 0.1 mm = 55.1 mm ± 0.18%

Mean thickness of one sheet = 55.1 mm/500 = 0.110 mm ± 0.18%

By measuring 500 sheets, the percentage uncertainty in the mean thickness of one sheet is reduced by a factor of 5.

This technique is particularly useful when applied to timing. By timing over multiple oscillations, the overall uncertainty will be reduced significantly and the effect of your reaction time will become less significant.

Knowledge check 41

a Why is the resolution of a stopclock (0.01 s) not used when you are using it to determine the uncertainty of timing an event taking about 3 s?

b Suggest a suitable uncertainty for a single such event.

Exam tip

The absolute uncertainty in a measurement made with a metre ruler is about 1 mm. Provided lengths in excess of 100 mm are measured, then the percentage uncertainty will be less than 1%.

Worked example

Figure 21 shows a stopwatch reading for 20 oscillations of a simple pendulum.

a What is the resolution of the instrument?

b Discuss what a realistic estimate of the uncertainty in this time would be.

c Explain why it is an advantage to time 20 oscillations when determining the period T of the oscillations.

Figure 21

d Determine a value for the period T and the estimated uncertainty in your value.

e Compare this uncertainty with that which would be incurred if only one oscillation had been timed.

Answer

a The resolution of the stopwatch is ±0.01 s.

b This is insignificant compared to human reaction time, which is at least 0.05 s. As this applies both when starting and stopping the timing, it would be appropriate to use ±0.1 s as the uncertainty in the measurement of time using a stopwatch.

c By measuring the time for 20 oscillations, and then dividing by 20, you would reduce the uncertainty by a factor of 20, as the error only occurs when the stopwatch is started and stopped.

d $20T = 12.60\,\text{s} \Rightarrow T = 0.630\,\text{s}$

Estimated uncertainty is reaction time of 0.1 s:

% uc in timing = $(0.1\,\text{s}/12.60\,\text{s}) \times 100\% = 0.8\% = \%$ uc in T

uncertainty in $T = 0.8\% \times 0.630\,\text{s} = 0.005\,\text{s} \Rightarrow T = 0.630 \pm 0.005\,\text{s}$

e If only one oscillation was timed, measured T would be $0.63 \pm 0.10\,\text{s}$:

% uc in $T = (0.1\,\text{s}/0.63\,\text{s}) \times 100\% = 16\%$

(This is 20 times as much as the percentage uncertainty when timing 20 oscillations.)

Using a set-square to reduce uncertainty

A set-square can be particularly useful to help determine precise measurements, as shown in Figure 22.

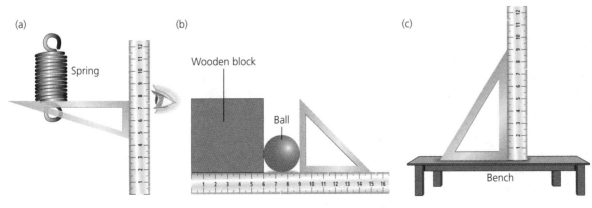

Figure 22

Exam tip

Always record the actual instrument reading and round down later.

Exam tip

Timing 20 oscillations and then dividing by 20 to find the time period reduces the percentage uncertainty by a factor of 20.

Knowledge check 42

A mass on a spring completes ten oscillations in 18.9 s. Determine the percentage uncertainty in the time period suggested by the precision of the recorded data.

Content Guidance

Make sure you keep your eye level, as shown in Figure 22(a). To make sure that a clamped ruler is vertical in relation to the bench (Figure 22c), a plumb-line can be used instead — a small metal bob on a piece of string will hang vertically if suspended freely. This can be clamped next to the ruler.

Knowledge check 43

Explain how you could use a set-square to check that a beam is horizontal with respect to the bench.

Worked example

Determining the spring constant of a spring

You are asked to investigate the limit of proportionality and determine the spring constant of a spring. Describe how you would do this. Your answer should include:
- a list of apparatus
- a diagram of the experimental set-up
- an outline of the procedure (including safety considerations)
- what you could do to reduce the uncertainty in your measurements of length and extension

Answer

Apparatus: retort stand and clamp, metre rule, set-square, slotted masses and hanger, spring

Set-up: Figure 23 shows how the apparatus is set up.

Figure 23

Exam tip

Note the use of a set-square as an aid to taking the reading.

Exam tip

Always measure an extension as the difference between two readings.

Safety considerations:
- Attach the masses securely to the spring and ensure that, should the masses fall, they will not fall on your feet. You may need to protect the floor with a crash pad.
- The clamp stand must be securely attached to the bench with a G-clamp.

Method:
- Secure the ruler as close to the spring as possible and use the set-square to ensure that the rule is vertical.
- Read the position of the bottom coil of the spring, using the set-square to ensure that the line between the spring and the ruler is horizontal.

→

- Align the eye with the top of the set-square to avoid parallax error.
- Add the mass hanger and masses one at a time and read the new position of the bottom ring each time until the extension is no longer linear.
- For each reading, calculate the extension of the spring by subtracting the original reading h_0 from the new reading h.

Using trigonometry to measure small angles

At best, a protractor can measure to 0.5° by interpolating between divisions. For an angle of, say, 5°, this would give rise to an uncertainty of 10%. By using trigonometric methods to measure small angles, we can significantly reduce this uncertainty. If you can measure the adjacent and opposite sides using a ruler, as shown in Figure 24, then the angle can be calculated using $\theta = \tan^{-1}$ (opposite/adjacent) = $\tan^{-1}(h/l)$.

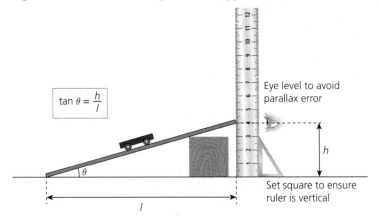

Figure 24

Knowledge check 44

a Calculate the angle of incline made by a 1.000 m ramp to the horizontal if the height of the raised end is 87 mm above the bench.

b Estimate the percentage uncertainty in your value for the angle.

Techniques for timing oscillations

Often, in experiments, the uncertainty in the reading is due not so much to the measuring instrument itself but rather to how it is used. There are a number of experiments at AS and A-level that require you to time an object that is oscillating.

Precautions when timing oscillations:

- Take care not to miscount oscillations — say 'nought' when starting to time.
- Use a *fiducial marker* (i.e. a pointer) behind the object to help you judge when the oscillating object passes the centre/equilibrium position (Figure 25). The speed of the oscillator is fastest as it passes through the equilibrium position, so you should time from this point. If you try to time from the maximum displacement, the object will be moving very slowly and it will be very difficult to judge when it has actually reached the point of maximum displacement and is about to reverse direction.

Exam tip

A stopwatch can measure to hundredths of a second, but is limited by the reaction time of the person using it. Always write down the reading on the stopwatch and then reduce this to an appropriate number of significant figures later.

Figure 25

Knowledge check 45

A pendulum bob is released a vertical distance of 10 cm above its equilibrium position. Calculate the maximum speed of the bob as it swings through the lowest point of its motion.

- Only start timing once the object has completed one or two oscillations. This will help you to start timing at the correct point. It will also ensure that you have not affected the motion of the object by giving it a small push at the start.
- Try to time at least 20 oscillations, if you can, in order to reduce the uncertainty. Obviously, if the system is heavily damped, then this will be difficult as the amplitude of the oscillations will decrease rapidly. In this case, take proportionately more repeat readings.

Worked example

Oscillations of a metre rule

A student makes some measurements of the time period T of the oscillations of a wooden metre rule in order to find a value for the Young modulus of the wood, using the arrangement shown in Figure 26. She finds in a textbook that the Young modulus E for the wood is given by the formula:

$$E = \frac{16\pi^2 ML^3}{wt^3 T^2}$$

where M is the mass suspended from the rule at a horizontal distance L from where the rule is clamped, and w and t are the width and thickness of the rule respectively.

Figure 26

Exam tip

You are expected to *comment on experimental design*, so, when writing a plan, always state the *reason* for your choice of each measuring instrument or for a particular technique that you use.

She plans to use:

- **two 100 g slotted masses** attached with elastic bands to either side of the rule, near the end, to make L as long as possible — with the slots in the masses perpendicular to the rule so that she can read the scale to determine the position of the centre of mass

→

- **a digital stopwatch** to time the oscillations — one that can record to a resolution of 0.01 s, which she considers to be more than adequate for timings that are likely to be of the order of 10 s
- **vernier callipers** to measure the width and thickness of the rule, at different points along its length — the callipers can read to a resolution of 0.1 mm
- a pin secured to the metre rule with a small piece of modelling clay to act as a **fiducial mark** to help her judge the centre of the oscillations

She records the following results:

$$M = 200\,g$$

$$L = 937 - 40 = 897\,mm$$

$$w = 28.5, 28.0, 28.6, 28.9\,mm;\ \text{average } 28.5\,mm$$

$$t = 6.6, 6.7, 6.6, 6.6\,mm;\ \text{average } 6.6\,mm$$

$$20T = 8.59, 8.54, 8.18, 8.61\,s;\ \text{average } 8.48\,s$$

a Explain why a value of 8.58 s for the average time should be used rather then the value stated by the student.

b Show that these data give a value of the order of 10^{10} Pa for the Young modulus of the wood.

c Explain the number of significant figures to which the final value should be given.

d Suggest a suitable uncertainty for the value of the period T.

e Discuss whether the selection of vernier callipers, which could only be read to a precision of 0.1 mm, to measure the width and thickness of the rule, was an *appropriate strategy*. Justify your answer by estimating the percentage uncertainty that these measurements introduce into the value obtained for the Young modulus.

f To what extent would using a micrometer screw gauge, or digital callipers, reading to 0.01 mm, reduce the percentage uncertainty in the value obtained for the Young modulus?

Answer

a Looking at the four values for $20T$, it would appear that the 8.18 reading is anomalous, as it is about 0.5 s smaller than the other values (suggesting only 19 oscillations may have been counted). This value should therefore be ignored (with a note to this effect) when determining the average value.

b Care needs to be exercised with units — all measurements need to be converted to SI units:

$$M = 0.200\,kg$$

$$L = 0.897\,m$$

$$w = 28.5 \times 10^{-3}\,m$$

→

$t = 6.6 \times 10^{-3}\,\text{m}$

$T = 8.58\,\text{s} \div 20 = 0.429\,\text{s}$

so

$$E = \frac{16\pi^2 M L^3}{wt^3 T^2} = \frac{16\pi^2 \times 0.200\,\text{kg} \times (0.897\,\text{m})^3}{28.5 \times 10^{-3}\,\text{m} \times (6.6 \times 10^{-3}\,\text{m})^3 \times (0.429\,\text{s})^2}$$

$$= 1.5 \times 10^{10}\,\text{Pa} \sim 10^{10}\,\text{Pa}$$

c The value should be quoted to two significant figures, $1.5 \times 10^{10}\,\text{Pa}$, because this is the number of significant figures of the least precise measurement — the thickness of the rule, 6.6 mm, has been measured to only 2 SF.

d Although the time taken for 20 oscillations has been recorded to a precision of 0.01 s, the range of values is $(8.61 - 8.54)\,\text{s} = 0.07\,\text{s}$. This is probably slightly less than the human reaction time. A more realistic uncertainty in $20T$ would be 0.1 s. This would give an uncertainty for the value of T of $0.1 \div 20 = 5\,\text{ms}$. We can then say $T = 0.429 \pm 0.005\,\text{s}$.

e The measurements recorded for the width were:

$w = 28.5, 28.0, 28.6, 28.9\,\text{mm}$; average 28.5 mm

The range of values is therefore $(28.0 - 28.9)\,\text{mm} = 0.9\,\text{mm}$.

If we use half the range (0.45 mm) as our uncertainty, we get:

$$\%\,\text{uc in } w = \frac{0.45\,\text{mm}}{28.5\,\text{mm}} \times 100\% = 1.6\%$$

Using a vernier scale reading to a precision of 0.1 mm to measure a thickness of 6.6 mm means that there is a percentage uncertainty in t of:

$$\frac{0.1\,\text{mm}}{6.6\,\text{mm}} \times 100\% = 1.5\%$$

As the term t^3 occurs in the formula for the Young modulus, the percentage uncertainty introduced by this measurement will be $3 \times 1.5\% = 4.5\%$.

The combined contribution of the measurements for width and thickness is therefore:

$\%\,\text{uc} = 1.6\% = 4.5\% = 6.1\%$

This is a not insignificant uncertainty. A vernier only reading to 0.1 mm to measure the thickness was not a good choice.

f If an instrument reading to 0.01 mm had been used, the percentage uncertainty in the thickness would have been reduced by a factor of 10, giving an uncertainty due to this measurement of less than 0.5%, which is perfectly acceptable. However, due to the relatively large spread of values in the measurement of the width, using a more sensitive instrument would not have had much effect on the previously calculated uncertainty of 1.6%. The overall uncertainty would therefore be about 2%, which is nevertheless a considerable improvement.

Using data loggers

Data loggers record data automatically and can record data with very small intervals or over long periods of time. They can record thousands of measurements per second, with the data output displayed on a computer as a graph or in a table. The number of readings taken per second is called the *sample rate*. Sensors are available to enable a wide range of measurements to be made, e.g. force, current, potential difference, magnetic flux density and temperature. The data logger is able to measure several variables simultaneously and so can, for example, record how force varies with time. This graph could then be used to calculate the impulse by determining the area underneath the line.

Knowledge check 47

Calculate the number of measurements taken in a minute by a data logger that is sampling at 0.5 Hz.

Worked example

In an experiment to investigate how the current changes with time as a capacitor is charged, a data logger attached to a current sensor is used instead of a digital ammeter and stopwatch. State three advantages of using a data logger in this way.

Answer

You could state:

- eliminates reaction time errors
- automatically takes current and time readings simultaneously
- updates the display much quicker than a standard ammeter
- high sample rate and so records many readings of current per second
- immediately measures the initial current the instant the experiment is started
- the data are stored in a computer and so can be easily processed and analysed (e.g. the gradient or area under a graph calculated)

Knowledge check 48

A magnet is dropped from rest through a vertical magnetic coil of length 50 mm. Show that it takes about 100 ms for the magnet to pass through the coil.

Worked example

A magnet is dropped through a coil of many turns. This is connected to a voltage sensor, which records the induced e.m.f. as the magnet falls through the coil (Figure 27a). Typical results are shown in Figure 27(b).

Figure 27

Discuss the advantages of using a data logger to collect these data.

Answer

You should include the following points:

- The event happens over about 100 ms, which does not give enough time for the data to be recorded manually.
- The high sample rate of the data logger enables the induced e.m.f to be measured hundreds of times during this short time interval.
- This produces enough data for a graph to be plotted showing how the induced e.m.f. varies with time.
- If the data logger is connected to a computer, the graph can be shown directly.

Analysing results

You will be expected to:

- *plot and interpret graphs*
- *process and analyse data using appropriate mathematical skills as exemplified in the mathematical appendix of the specification*

Working through the *Maths and units* section of this student guide will give you plenty of practice in developing your mathematical skills. You should note that more advanced mathematical skills are required for A-level than for AS, which mainly means ensuring that you are confident in handling exponential functions and drawing and analysing logarithmic graphs.

Plotting a graph is a key first step in analysing your results. All graphs should be drawn with sensible scales and with appropriately labelled axes, including correct units. Points should be plotted accurately and a thin straight line, or smooth curve, of best fit drawn.

What graph should I draw?

Physicists often carry out experiments in order to find an equation that relates the independent (x) and dependent (y) variables. If the graph obtained is a straight line of gradient m and a y-intercept of c, the relationship can be written in the form of $y = mx + c$.

If the graph is a straight line *passing through the origin* ($c = 0$), then we can say that variable y is directly proportional to variable x ($y = mx$). For example, in an experiment to measure the resistance of a wire, the graph shown in Figure 28 was obtained.

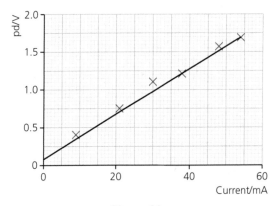

Figure 28

The graph in Figure 28 does not quite pass through the origin, which is probably due to a systematic error in the measurements. This could be due to a zero error in the ammeter or voltmeter, or possibly some contact resistance at the terminals where the wire is connected into the circuit.

If the line is a straight line but clearly does not pass through the origin, then we can say that there is a *linear relationship* between the two variables, but it is *not* a proportional relationship.

Many experiments result in graphs that are curves. Perhaps there is an inversely proportional relationship between the variables, or a power law? Maybe the relationship is exponential?

Exam tip

You will lose marks if the line is forced through the origin or if a straight line is drawn when the data suggest a curve.

Worked example

Oscillations of a metre rule

The student's teacher in the example on page 36 suggests that a better value for the Young modulus might be obtained if measurements were taken with different masses and a suitable graph was plotted. The relevant equation is:

$$E = \frac{16\pi^2 ML^3}{wt^3 T^2}$$

a For a given metre rule, this equation for the Young modulus contains three variables. State what they are and, following the teacher's suggestion, state which one must be kept constant. How would you ensure this?

b Explain what should be plotted in order to get a linear graph.

c Show how the graph could be used, together with the rest of the experimental data, to get a value for the Young modulus.

d Suggest why a graph is likely to reduce the uncertainty in the value obtained for the Young modulus.

Answer

a For a given rule, the three variables are:
- the suspended mass M
- the period T
- the length of suspension L

Of these, L should be kept constant and T should be found for different values of M.

To ensure L remains constant:
- the masses should be secured tightly
- the distance should be checked before each timing

b From the equation:

$$E = \frac{16\pi^2 ML^3}{wt^3 T^2} \Rightarrow T^2 = \frac{16\pi^2 L^3}{Ewt^3} \times M$$

Therefore, a graph of T^2 against M should be plotted.

Exam tip

Practise so that you are confident in rearranging complex formulae to give straight-line graphs of the form $y = mx + c$ and in determining which of the terms are variables and which are constant.

c The graph should be a straight line through the origin:

$$\text{gradient} = \frac{16\pi^2 L^3}{Ewt^3} \Rightarrow E = \frac{16\pi^2 L^3}{\text{gradient} \times wt^3}$$

d Plotting a graph would:
- give an average of a number of readings, and so
- reduce both random and systematic errors

Lines of best fit

Before you draw a line of best fit, think about the underlying physics to help you decide whether the line should be straight or curved. Perhaps there is a law that applies to these data (e.g. $V = IR$)?

Use these guidelines to ensure that you draw the best possible line of best fit. Examiners expect lines to be drawn carefully (i.e. thin and straight or smooth).
- Try to make sure there are as many points on one side of the line as the other.
- The line should pass very close to the majority of plotted points.
- Are there any obviously anomalous results?
- Are there uncertainties in the measurements? The line of best fit should fall within error bars if those are drawn.
- Use a thin, sharp pencil to draw the line of best fit.

Worked example

How *not* to plot graphs!

Look at the graphs in Figure 29 and decide what is wrong with each one.

(a)

(b)

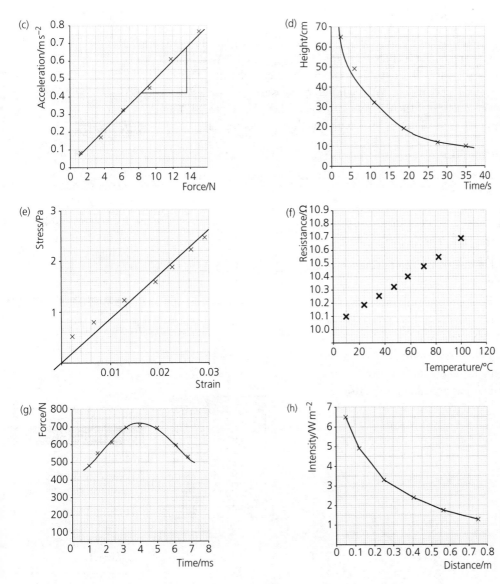

Figure 29

Answer

a Awkward scale divisions — in particular, avoid multiples of 3 as interpolation is difficult.

b No readings between 0.1 A and 0.5 A — a poor distribution of readings makes it difficult to plot a line of best fit.

c The triangle used to calculate the gradient is too small — this would produce large percentage uncertainties in the values of Δx and Δy and therefore in the gradient.

d Poorly drawn curve of best fit — a smooth curve is required (this line is 'spidery').

e The line has been forced through the origin, and so is not the 'best fit'.

f A sharp pencil was not used to plot points, making them very thick — precision is lost.

g Poor y-scale — the points do not fill more than half the y-axis, so the scale could have been doubled.

h Straight lines have been drawn joining the points — a *smooth* curve of best fit should have been drawn.

Dealing with anomalous results

You may find that some results do not fit the general pattern and lie a long way from the line of best fit. Circle these and do not include them in your line of best fit. You should make it clear that you have done this. Try to think what may have caused them. Did a different student collect that piece of data? Was a different measuring device used? Did the external conditions change? If there is time, this point should be repeated. You should also exclude any anomalous values when calculating the mean of repeats.

Calculating gradients

A gradient often represents an important quantity, as discussed on page 13 in the section on mathematical requirements.

Worked example

Determining the resistance of a component

The circuit shown in Figure 30 is set up to determine the resistance of a component.

The p.d. (V) across the component and the current (I) in it were measured for a range of voltages and the results shown in Table 7 were obtained.

a Plot a graph of V on the y-axis against I on the x-axis.

b Comment on the graph you obtain.

c Use your graph to determine a value for the resistance of the component.

Answer

Figure 30

Table 7

I/µA	V/mV
5.0	1.75
10.0	1.27
15.0	5.10
20.0	6.72
25.0	8.35
30.0	9.88

a
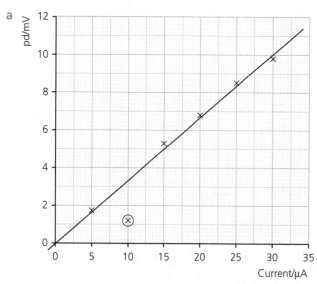

Figure 31

b The graph is a straight line through the origin. The point (10.0 µA, 1.27 mV) is an anomalous result, and so is excluded from the line of best fit.

c $R = V/I = \text{gradient} = (10.0 - 0.0) \times 10^{-3}/(30.0 - 0.0) \times 10^{-6}$
   $R = 333\,\Omega = 330\,\Omega$ (2 sf)

Worked example

Calculating the spring constant

Explain how you would use the data for the spring constant experiment on page 34 to determine the limit of proportionality and the spring constant.

Answer

- For each extension, calculate the force applied using $W = mg$.
- Plot a graph of force against extension. Note that, contrary to normal practice, we are plotting the dependent variable (extension) on the x-axis. This is so that the gradient gives us the spring constant.
- Read off the point at which the graph begins to curve — this is the limit of proportionality.
- For the linear region, determine the gradient by drawing a large triangle — this gives the spring constant.

Error bars and uncertainties in the gradient

The uncertainty in a measurement can be shown on a graph as an error bar. If there is an uncertainty in both the quantities y and x, then instead of an error bar you would have an error rectangle.

If you are plotting a log graph, then you have to calculate the logarithm of the maximum and minimum possible values for each data point and determine the error bars using these values.

The uncertainty in each measurement can be determined from the resolution of the data, or it is sometimes quoted as a percentage. Use the following guidelines to help you draw correct error bars.

If there are repeat readings:

- plot the data points at the mean values
- calculate the range of the data, ignoring any anomalies
- add error bars with lengths equal to half the range on either side of the data points
- if you know the absolute uncertainty, plot the error bars based on this straight away

Exam tip

When plotting graphs, remember to plot the correct quantities on each axis. *Force* against *time* means that *force* would go on the y-axis and *time* on the x-axis.

Knowledge check 53

Determine the uncertainty in ln T that should be used to plot an error bar if $T = (8.2 \pm 0.2)$ N.

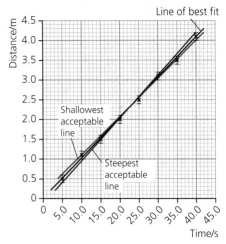

Figure 32

To determine the uncertainty in a gradient, three lines should be drawn on the graph, as shown in Figure 32:

- the line of 'best fit' and
- the steepest (shown in blue) *and* the shallowest (shown in red) gradient lines that can be drawn through the data points (or through the error bars if these are shown)

The gradient of each line should be found, then:

$$\text{percentage uncertainty in gradient} = \frac{(\text{highest possible gradient} - \text{lowest possible gradient})}{\text{gradient of line of best fit}} \times 100\%$$

The percentage uncertainty in the *y*-intercept can be found in a similar way.

Exponential changes and log graphs

(A-level only) At A-level you will need to be familiar with exponential changes. Common examples include radioactive decay, the decrease in amplitude of a damped oscillator and the discharge of a capacitor through a resistor.

A good example of an exponential relationship is shown in an experiment to investigate how the length of a jelly fibre affects the amount of light that is absorbed (Figure 33). A phototransistor is used to detect the light. This produces a voltage that is directly proportional to the amount of light falling on it.

Figure 33

A graph of the data obtained in such an experiment is shown in Figure 34.

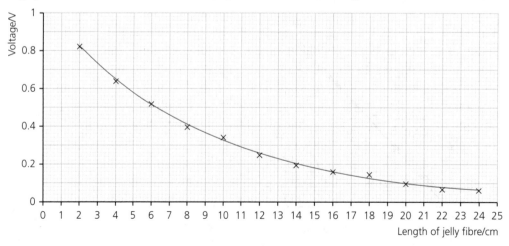

Figure 34

The exponential shape of the curve suggests that the intensity of light passing through the fibre might decrease exponentially with length according to the relationship:

$$I = I_0 e^{-\mu x}$$

where I = intensity of light, μ = absorption coefficient of jelly and x = length of jelly.

Taking natural logarithms ('ln') on both sides of the equation gives:

$$\ln I = \ln I_0 - \mu x$$

Since the phototransistor voltage V is proportional to the intensity I:

$$\ln V = \ln V_0 - \mu x$$

Comparing with $y = mx + c$, we can see that, if $\ln V$ is plotted on the y-axis against x on the x-axis, we should expect a straight-line graph of gradient = $-\mu$ and y-intercept = $\ln V_0$ as shown in Figure 35. This is called a log-linear graph.

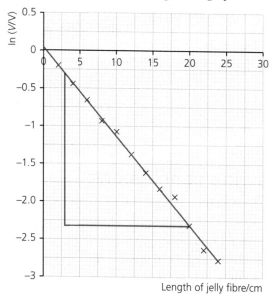

Figure 35

Exam tip

You can also test for exponential changes mathematically using the 'constant ratio test'. Applying this to the data shown in Figure 34:

$0.64/0.82 \approx 0.78$

$0.52/0.64 \approx 0.81$

$0.40/0.52 \approx 0.77$

These ratios are all within a few per cent of each other, so within experimental limits we can say that the data show an exponential decrease in voltage (light intensity) with an increase in length.

Worked example

a Use the graph in Figure 35 to determine the absorption coefficient, μ.

b Use this graph to find the value of V_0.

c Determine the length of fibre needed to reduce the intensity to 15% of its initial amount.

Answer

a Taking the points (3.00, −0.30) and (20.0, −2.30) gives:

$$\text{gradient} = \frac{-2.30 - (-0.30)}{(20.0 - 3.0)\,\text{cm}} = -0.12\,\text{cm}^{-1}$$

The absorption coefficient is therefore $0.12\,\text{cm}^{-1}$.

b y-intercept = $0.04 \Rightarrow V_0 = e^{0.04} = 1.0\,\text{V}$

Exam tip

Always include the units when determining the gradient of any graph.

→

c Substituting $I = 0.15I_0$ and $\mu = 0.12\,\text{cm}^{-1}$ into the equation $I = I_0\,e^{-\mu x}$:

$$0.15I_0 = I_0\,e^{-0.12x} \Rightarrow \ln(0.15) = -0.12(x/\text{cm}) = -1.897$$

$$(x/\text{cm}) = -1.897/(-0.12)$$

$$x = 16\,\text{cm}$$

Exam tip

Always try to check the validity of your answer if possible. Here you should look back at Figure 34 and see whether the voltage has dropped to 15% of its original value for a length of 16 cm.

Log graphs can also be used to investigate relationships of the form:

$$y = Ax^n$$

where A and n are constants.

It would take a long time to find n using trial and error, but if we take logs of both sides (either to base '10' or to base 'e') the equation becomes:

$$\log y = \log A + n\log x \qquad \text{or} \qquad \ln y = \ln A + n\ln x$$

A graph of $\ln y$ against $\ln x$ will give a straight line. Comparing this with the equation of a straight line, $y = mx + c$, gives:

$$\ln y = \ln A + n\ln x$$
$$y = c + mx$$

i.e. gradient $m = n$ and intercept $c = \ln A$.

Exam tip

For an exponential function, you *must* use natural logs ('ln'). For investigating a power relationship, you can use either 'ln' or 'log'. If you always use 'ln', you cannot go wrong.

Evaluating results and drawing conclusions

You will be expected to *evaluate results and draw conclusions with reference to measurement uncertainties and errors*.

Exam questions may ask you to discuss to what extent your results confirm a particular relationship between two variables. You may also be asked to comment on accuracy and precision, and to discuss how the errors and uncertainties in the experiment could be reduced. The following worked example gives you practice at applying your knowledge of *planning, making measurements, recording data, analysing results* together with *evaluating results and drawing conclusions*.

Exam tip

The y-intercept is $\ln A$ (or $\log A$) and so $A = e^{y\text{-intercept}}$ (or $10^{y\text{-intercept}}$).

Exam tip

Reminder: the percentage difference between two experimental values is given by:

$$\%\text{difference} = \frac{\text{difference between the values}}{\text{average of the two values}} \times 100\%$$

The percentage difference between an experimental value and a stated or known value is given by:

$$\%\text{difference} = \frac{\text{difference between the values}}{\text{stated value}} \times 100\%$$

Worked example

Investigating resonance

(A-level only) A student plans to investigate how the frequency of vibration of air in a conical flask depends on the volume of air in the flask. She blows directly into the neck of the flask and listens to the sound of the air vibrating. She then pours water into the flask until it is approximately half filled and blows into the flask as before. She notices that the pitch (frequency) of the vibrating air is higher when the flask is half full of water.

She thinks that there might be a relationship between the natural frequency of vibration f of the air in the flask and the volume V of air in the flask, of the form $f \propto V^n$, where n is a numerical constant.

The student sets up the arrangement shown in Figure 36.

Figure 36

With the flask empty, she increases the frequency of the signal generator until the air in the flask vibrates very loudly. She repeats this for increasing volumes of water in the flask.

a Explain why this happens at the natural frequency of vibration of the air.

b Describe a technique by which the uncertainty in determining this frequency could be reduced.

c Describe how she could vary, and measure, the volume of air in the flask.

d Explain how plotting a graph of $\ln f$ against $\ln V$ would enable her to test whether $f \propto V^n$ and would enable her to find a value for n.

e The student obtained the data shown in the table.

V/cm^3	f_1/Hz	f_2/Hz	f/Hz
554	217	221	219
454	243	241	242
354	271	277	274
254	325	323	324
204	363	359	361
154	412	418	415

Calculate values of $\ln(V/cm)$ and $\ln(f/Hz)$ from the data in the table and add those values to the table. Then plot a graph of $\ln(f/Hz)$ against $\ln(V/cm)$ and use your graph to determine a value for n.

f Explain **qualitatively** whether your value for n is consistent with the student's initial observations.

Answer

a When the frequency of the signal generator, i.e. the frequency of vibration of the loudspeaker, is equal to the natural frequency of vibration of the air inside the flask, resonance occurs. Energy is transferred with maximum efficiency from the loudspeaker to the air in the flask, and so this air vibrates very loudly.

b The frequency should be increased gradually until resonance occurs. This frequency should be recorded. The frequency should then be increased beyond the resonant frequency and gradually reduced until once again resonance is detected. This frequency should also be recorded and the average of the two values should be taken as the resonant frequency.

c A measuring cylinder is required, then:
- this is used to fill the flask to the top with water, so that the volume V_f of the flask can be determined
- the flask is then emptied
- a known volume v of water is now added to the flask
- the volume of air will therefore be $V = V_f - v$
- further volumes of water are added to give different volumes V of air in the flask

d The proposed equation is $f \propto V^n$, or $f = kV^n$. If we take logarithms ('ln') on both sides of the equation, we get:

$$\ln(f) = n\ln(V) + \ln(k)$$

So, if a graph of $\ln(f)$ against $\ln(V)$ is plotted, this should be a straight line with gradient equal to the constant n and intercept on the y-axis equal to $\ln(k)$.

e The values are shown in the table.

V/cm^3	f/Hz	$\ln(V/cm^3)$	$\ln(f/Hz)$
554	219	6.32	5.39
454	242	6.12	5.49
354	274	5.87	5.61
254	324	5.54	5.78
204	361	5.32	5.89
154	415	5.04	6.03

→

> **Exam tip**
>
> Note that the technique should be described in detail — an answer such as 'The reading was repeated and the average taken' is not sufficient. This technique can also be applied to focusing an image with a lens.

Your graph should be a straight line of negative slope, as in Figure 37.

Figure 37

Note that:

- the points should occupy at least half the graph paper in each direction, which means that the scales should *not* start at the origin
- the axes should be labelled exactly as shown: $\ln(f/\text{Hz})$ on the *y*-axis and $\ln(V/\text{cm})$ on the *x*-axis
- the points should be plotted to a precision of at least half a square
- a thin, straight line of best fit should then be drawn

The constant *n* is the gradient of the graph. Taking a large triangle:

$$\text{gradient} = n = \frac{6.06 - 5.36}{5.00 - 6.40} = \frac{+0.70}{-1.40} = -0.50$$

(do not forget the minus sign)

f A negative value of *n* means that, as the volume of air in the flask gets smaller, the frequency of vibration (pitch) of the air gets higher, which is consistent with the student's initial observation.

A value of −0.50 suggests that $n = -\frac{1}{2}$, meaning that:

$$f \propto V^{-1/2} \quad \text{or} \quad f \propto \frac{1}{\sqrt{V}}$$

In other words, the frequency of vibration is inversely proportional to the square root of the volume of air.

Knowledge check 56

(A-level only) The period *T* of vertical oscillations of a mass *m* on a spring of spring constant *k* is given by:

$$T = 2\pi\sqrt{\frac{m}{k}}$$

a Write down the expression obtained by taking natural logarithms on both sides of this equation.
b State the value of the gradient when $\ln(T/\text{s})$ is plotted against $\ln(m/\text{kg})$.
c The *y*-intercept is found to be 0.19. What value does this give for the spring constant *k*?

Core practicals

Core practical 1

Determining the acceleration of a free-falling object

Essential theory

A free-falling object means one that is falling vertically under gravity, with no other forces acting. In practice, an object falling in air will experience air resistance. But provided the object is made of a dense material and its speed is not excessive, it may be considered to be falling freely. We can then apply the equations of uniformly accelerated motion to the object — in particular:

$$s = ut + \tfrac{1}{2}at^2$$

If the object is released from rest, $u = 0$. If the object then falls through a height h with acceleration g we get:

$$h = \tfrac{1}{2}gt^2$$

Rearranging gives us:

$$t^2 = \frac{2}{g} \times h$$

If we plot a graph of t^2 on the y-axis against h on the x-axis, the gradient will be $2/g$.

Planning

There are several ways of doing this experiment: using light gates with a computer interface, strobe photography and video-frame analysis are examples. But, in all cases, g is calculated from the time taken for a free-falling object to move through a measured distance. The main problem is the relatively short times involved.

Knowledge check 57

Show that the time taken for an object to fall 1.2 m under gravity is about 0.5 s.

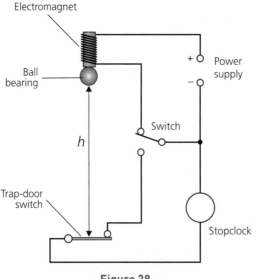

Figure 38

Figure 38 shows one possible method. The points to remember are as follows:

- The time is measured using an electronic stopclock that reads to 1/100th of a second.
- When the switch is in the position shown, the electromagnet is activated and holds the steel ball bearing in place.
- When the switch is thrown, the electromagnet circuit is broken, releasing the ball, and simultaneously the clock circuit is activated, starting the clock.
- The clock is stopped when the ball hits the trapdoor and breaks the circuit.
- The time should be determined at least three times and averaged for each height, over as wide a range of heights as possible.

Safety

The following precautions should be observed:

- Although low voltages are being used, care should be taken with wiring to prevent any short circuits.
- A suitable container should be placed under the trapdoor to catch the ball bearing to prevent it falling on the floor and creating a trip hazard.

Worked example

Table 8 shows a typical set of data. In order to simplify this example, the times indicated are the average of three values for each height. In practice *all* the values should be shown.

Table 8

h/m	t/s	t^2/s^2
0.600	0.38	
0.800	0.44	
1.000	0.47	
1.200	0.52	
1.400	0.56	
1.600	0.59	

Exam tip

You should always record and show *all* your readings.

a Complete Table 8 to show the values of t^2.

b Plot a graph of t^2 against h.

c Determine a value for g from the gradient of your graph.

d Estimate the uncertainty in your value for g, with an explanation of your reasoning.

e Suggest why the graph has a small positive intercept on the t^2 axis.

Answer

a The values of t^2/s^2 are: 0.14, 0.19, 0.22, 0.27, 0.31 and 0.35, respectively.

b Your graph should be like Figure 39:

$\rightarrow$

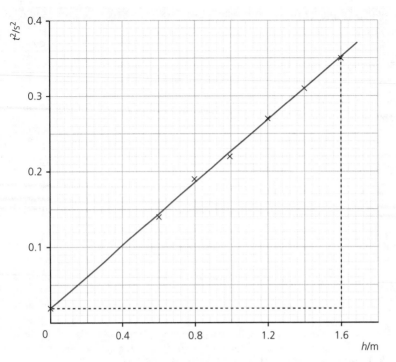

Figure 39

Exam tip

You should always draw the line of best fit through the experimental points and not force it through the origin, as there may be a systematic error. One of the reasons for drawing a graph is to identify, and make allowance for, such errors.

Exam tip

Remember to draw a large triangle when determining a gradient and to show the coordinates you have taken in your calculation.

c gradient $= \dfrac{(0.35 - 0.02)\,\text{s}^2}{(1.60 - 0.00)\,\text{m}} = 0.20(6)\,\text{s}^2\text{m}^{-1} = \dfrac{2}{g}$

This gives g $= \dfrac{2}{0.20(6)\,\text{s}^2\text{m}^{-1}} = 9.7\,\text{m s}^{-2}$

d As the values of t could only be measured to 2 SF, it would be reasonable to state the value for g as $9.7 \pm 0.1\,\text{m s}^{-2}$. Without all the timings, a more detailed analysis is not possible.

e A small positive intercept on the t^2 axis suggests a systematic error, either in t (e.g. a delay in releasing the ball after the switch has been thrown or a delay in the trapdoor opening after the ball has hit it) or in the measurement of h (all the h values being too small).

Core practical 2

Determining the electrical resistivity of a material

Essential theory

The resistivity ρ of a material is given by the equation:

$$R = \frac{\rho l}{A}$$

where R is the resistance of a length l of the material, having area of cross-section A.

If we plot a graph of R on the y-axis against l on the x-axis, we should get a straight line through the origin of gradient ρ/A.

If we determine A for the sample, we can then find ρ from $\rho = \text{gradient} \times A$.

Planning

The easiest method is to find the resistance of different lengths of resistance wire, such as nichrome, and measure the diameter d of the wire to determine A from $A = \pi d^2/4$.

Note that:

- As the resistance of a wire such as this is only a few ohms, an ohmmeter is unsuitable for measuring the resistance with any degree of precision.
- Therefore, an ammeter and a voltmeter must be used, and the resistance determined from $R = V/I$.
- The circuit should be set up as shown in Figure 40(a).
- The movable crocodile clip should be pressed firmly on to the wire to minimise contact resistance (but not too firmly or the wire will be damaged).

(a)

(b)

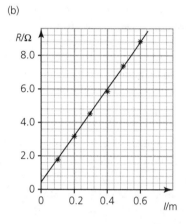

Figure 40

Safety

The following precautions should be observed:

- Although low voltages are being used, care should be taken with wiring to prevent any short circuits.
- See also question **h** in the worked example below.

a State what instrument you would use to measure the diameter d of the wire. Justify your choice of instrument.

b The following readings were recorded for the diameter of the wire:
d/mm = 0.31, 0.33, 0.30, 0.39, 0.32
Explain what value you would take for the average diameter.

c Estimate the percentage uncertainty in your value for the diameter of the wire.

d Figure 40(b) shows how the resistance R varies with the length l for such a wire. The intercept on the resistance axis indicates a systematic error. Suggest a possible reason for this.

e Determine the gradient of the graph and hence find a value for the resistivity of nichrome.

→

f Assuming that the other experimental uncertainties are much less than the uncertainty you calculated in c, estimate the uncertainty in your value for the resistivity.

g The cell has an e.m.f. of 1.5 V. Show that the current in the wire is about 0.5 A when a length of 20 cm is connected into the circuit.

h Explain the problem that could arise when such a short length of wire is connected. Suggest any additions that you could make to the circuit to overcome this problem.

Answer

a A micrometer (or digital vernier) would be suitable, as it could measure a diameter of about 0.3 mm with adequate precision (0.01 mm or possibly 0.001 mm, depending on the instrument).

b The 0.39 mm value is obviously a spurious reading and so should be ignored when calculating the average. The average is therefore:

$$d = (0.31 + 0.33 + 0.30 + 0.32)\,\text{mm} \div 4 = 0.31(5)\,\text{mm}$$

c The uncertainty in d should be taken as half the range of values, which is 0.015 mm. The percentage uncertainty in d is therefore:

$$\%\,\text{uc in}\,d = \frac{0.015\,\text{mm}}{0.315\,\text{mm}} \times 100\% = 5\%$$

d The intercept indicates a systematic error — a small resistance for zero length of wire! This is probably caused by a poor contact between the crocodile clip and the wire at the zero end, giving rise to a small contact resistance. Although there will also be contact resistance between the movable clip and the wire, this will depend on how hard the clip is pressed against the wire, and so is more likely to give rise to random errors.

e gradient $= \dfrac{(8.80 - 0.40)\,\Omega}{(0.60 - 0.00)\,\text{m}} = 14.0\,\Omega\,\text{m}^{-1}$

$A = \pi d^2/4 = [\pi \times (0.315 \times 10^{-3}\,\text{m})^2] \div 4 = 7.7(9) \times 10^{-8}\,\text{m}^2$

$\rho = \text{gradient} \times A = 14.0\,\Omega\,\text{m}^{-1} \times 7.7(9) \times 10^{-8}\,\text{m}^2 = 1.09 \times 10^{-6}\,\Omega\,\text{m}$

f Percentage uncertainty in ρ = percentage uncertainty in A
As the area is $\pi d^2/4$, the percentage uncertainty in A is $2 \times 5\% = 10\%$.
Uncertainty in ρ = 10% of $1.09 \times 10^{-6}\,\Omega\,\text{m} = 0.109 \times 10^{-6}\,\Omega\,\text{m}$.
The value of ρ is therefore best expressed as $(1.09 \pm 0.11) \times 10^{-6}\,\Omega\,\text{m}$.
[Note that, as there are several values for the diameter, enabling us to get an average with an uncertainty calculated from the spread of readings, we are justified (just) in giving the value for ρ to 3 SF. With a single reading for the diameter, only 2 SF could be justified.]

g The resistance of a 20 cm length of wire is $0.2\,\text{m} \times 14.0\,\Omega\,\text{m}^{-1} = 2.80\,\Omega$

$$I = \frac{V}{R} = \frac{1.5\,\text{V}}{2.80\,\Omega} = 0.54\,\text{A} \approx 0.5\,\text{A}$$

h A current of the order of 0.5 A could heat the wire and therefore change its resistivity. This could be overcome by adding a resistor (e.g. 10 Ω) in series with the cell and ammeter.

> **Exam tip**
> Remember that, if a quantity is squared, then its percentage uncertainty is doubled.

Core practical 3

Determining the e.m.f. and internal resistance of an electrical cell

Essential theory

When there is a current I in an electric cell, the terminal p.d. V is less than the e.m.f. ε of the cell because work has to be done overcoming the internal resistance r of the cell. This can be expressed by the equation:

$$V = \varepsilon - Ir$$

or rearranging:

$$V = -rI + \varepsilon$$

If we measure the p.d. for different values of current in the cell, a graph of V on the y-axis against I on the x-axis will be a straight line of gradient $-r$ and intercept ε (provided the internal resistance is constant).

The simplest experiment is to do this for a dry cell, such as a 1.5 V AA cell. An interesting alternative is to investigate what happens with a solar cell.

Planning

The circuit required is shown in Figure 41.

- The solar cell is illuminated with a bench lamp (not shown in Figure 41).
- The current in the cell is adjusted with the variable resistor and the corresponding p.d. recorded.

Safety

The following precautions should be observed:

- Although low voltages are being used, care should be taken with wiring to prevent any short circuits.
- Digital meters should be handled with care to prevent damage.
- Be careful not to touch the lamp — it is likely to be very hot (unless it is an LED).

A high resistance voltmeter is connected across an AA cell and reads 1.58 V. A 4.7 Ω resistor is then connected across the cell and the voltmeter reading drops to 1.50 V. Explain why the voltage drops, and use the data to calculate the internal resistance of the cell.

Figure 41

Exam tip

Remember to distinguish between a solar *panel*, which converts radiation from the Sun into *thermal* energy by heating water circulating in an array of pipes, and a solar *cell*, which converts radiation from the Sun into *electrical* energy. An array of solar cells is also commonly called a solar panel.

Worked example

A typical set of results is shown in Table 9.

Table 9

I/mA	0.100	0.200	0.300	0.400	0.500	0.600	0.700	0.800	1.000	1.200
V/V	2.63	2.61	2.54	2.52	2.49	2.44	2.38	2.33	2.14	1.70

a A 47 kΩ rotary potentiometer was used for the variable resistor. Use the data in Table 9 to explain why this is a suitable choice.

b What is the control variable that must be kept constant in this experiment?

c Explain why you should keep well away from the cell when taking readings.

d Plot a graph of V/V on the y-axis against I/mA on the x-axis.

e Explain the shape of your graph. Suggest why there is some scatter in the points.

f Use your graph to find the e.m.f. of the solar cell and its internal resistance for current values of less than 600 μA.

→

Answer

a When $I = 1.200\,\text{mA}$, $V = 1.70\,\text{V}$, so $R = V/I = 1.70\,\text{V}/1.200\,\text{mA} = 1.4\,\text{k}\Omega$.
  When $I = 0.100\,\text{mA}$, $V = 2.63\,\text{V}$, so $R = V/I = 2.63\,\text{V}/0.100\,\text{mA} = 26\,\text{k}\Omega$.
  The resistance needs to have a range from $1.4\,\text{k}\Omega$ to $26\,\text{k}\Omega$. Therefore a $47\,\text{k}\Omega$ rotary potentiometer, which can be varied from 0 to $47\,\text{k}\Omega$, will be suitable.

b As the voltage from a photocell depends on the illumination, the intensity of the light falling on it is the control variable and must be kept constant. This means that the position of the lamp must not be altered during the course of the experiment.

c You should keep well away from the photocell when taking readings to make sure that you do not get in the way of the light falling on the cell.

d Your graph should look like Figure 42.

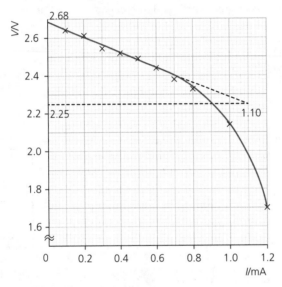

Figure 42

e The graph is initially a straight line of negative slope ($= -r$) indicating a constant internal resistance r up to about $600\,\mu\text{A}$. After $600\,\mu\text{A}$ the graph curves downwards, indicating that the internal resistance is getting larger (because the p.d. is decreasing more than it would do if r was constant). The scatter in the points is probably due to slight changes in illumination.

f e.m.f.: ε = intercept on V axis when $I = 0 \Rightarrow \varepsilon = 2.68\,\text{V}$
  The internal resistance for current values of less than $600\,\mu\text{A}$ is given by the gradient of the *linear part* of the graph. Note that this has been extended to give a *large* triangle.

$$r = -\text{gradient} = -\frac{(2.68 - 2.25)\,\text{V}}{(0.00 - 1.10)\,\text{mA}}$$

$$= 390\,\Omega \text{ to 2 SF}$$

Exam tip

Ideally, the intensity of the light falling on the photocell should be checked during the experiment with a light meter to ensure that it remains constant. The intensity of the light under which the experiment was conducted should then be recorded.

Exam tip

Note that the scale on the y-axis of the graph should *not* start at $V = 0$ or the scale would be far too small. Your plotted points should occupy *at least* half the space available. Also, the straight part of the line has been extended to give a large triangle for the gradient.

Knowledge check 61

When the current in the solar cell is $1.20\,\text{mA}$, the p.d. across its terminals is $1.70\,\text{V}$. Show that this gives a value for the internal resistance of about $800\,\Omega$ (see answer **e** in the worked example).

Core practical 4

Using a falling-ball method to determine the viscosity of a liquid

Essential theory

The terminal velocity v of a sphere of radius r and density ρ_s falling through a liquid of density ρ_l and viscosity η is given by:

$$v = \frac{2(\rho_s - \rho_l)gr^2}{9\eta}$$

If we determine the terminal velocity v for spheres of the same material (e.g. steel ball bearings) having different radii r and plot a graph of v on the y-axis against r^2 on the x-axis, the gradient will be $\dfrac{2(\rho_s - \rho_l)g}{9\eta}$.

If ρ_s and ρ_l are then determined, a value for η can be found.

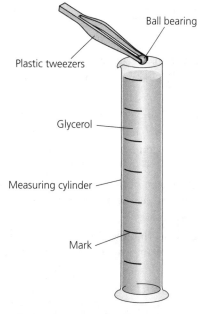

Figure 43

Planning

The following points should be noted:

- To prevent having to use a very tall cylinder, it is necessary for the spheres to reach terminal velocity soon after entering the liquid.
- This can be achieved by using small spheres of a dense material (e.g. steel ball bearings) and a fairly viscous liquid (e.g. glycerol).
- The spheres are dropped carefully into the liquid using plastic tweezers, as shown in Figure 43.
- The terminal velocity is determined by timing how long it takes each sphere to fall between two marks on the cylinder that are a known measured distance apart.

Safety

The following precautions should be observed:

- Take care to avoid contact with the glycerol, as you may be allergic to it — it would be a good idea to wear disposable gloves.
- Use a magnet to retrieve the ball bearings and clean them with a paper towel.
- Ensure that the cylinder is stable and does not topple over — use a clamp if necessary.

Knowledge check 62

Sketch a graph of velocity against time for a sphere falling through a fluid. Use your graph to explain what is meant by *terminal velocity*.

Knowledge check 63

Use the terminal velocity equation to show that the units of viscosity are Pa s.

Knowledge check 64

Show that the mass of a steel ball bearing of diameter 3 mm is about 0.1 g. The density of steel is 7800 kg m^{-3}.

Content Guidance

Worked example

Table 10 shows a set of measurements recorded by a student.

Table 10

Diameter d/mm	r^2/mm^2	Time/s	Average time t/s	Terminal velocity
2.53	1.6	12.64, 12.52	12.58	1.59
3.02	2.28	8.81, 8.89	8.85	2.26
4.00	4	4.99, 4.91	4.95	4.04
5.01	6.275	3.97, 4.09	4.03	6.30
5.96	8.88	2.31, 2.21	2.26	8.84
Distance between marks 20 cm				

a Comment on the presentation of these readings.

b Explain how you would check that terminal velocity had been reached.

c Describe how you would determine the density of the ball bearings and the density of the glycerol.

d Plot a graph of v/m s^{-1} against r^2/m^2.

e The density of steel is 7800 kg m^{-3} and the density of glycerol is 1260 kg m^{-3}. Determine the gradient of your graph and hence a value for the viscosity η of glycerol.

f Discuss whether the measurement of the diameter, or the determination of the terminal velocity, is likely to give rise to the greater experimental uncertainty in the value obtained for the viscosity.

g The student's teacher comments that the student has not recorded the temperature of the glycerol. Explain why recording the temperature is important.

Answer

a In the table:
- There are no units for the terminal velocity — from the values it would appear that the unit should be cm s^{-1}.
- There is inconsistency in the number of significant figures for values of r^2 — from the readings recorded for d, the values of r^2 should all be expressed to 3 SF.
- There are no repeat readings shown for d and more than five values of d would be better for drawing a graph.
- The distance of fall '20 cm' has only been given to the nearest cm when mm precision should have been possible.

b Whether terminal velocity has been reached can be checked by measuring the time (using the lap timer facility on the stopwatch) to fall between successive marks. If the times are equal (within experimental uncertainty), then terminal velocity has been reached.

c As the mass of a 3 mm sphere is only about 0.1 g, a sensitive balance (0.01 g or better) is needed to find the mass. The mass m of each sphere should be found and then the density determined from:

$$\rho_s = \frac{m}{\frac{4}{3}\pi r^3}$$

$\rightarrow$

The density of the glycerol can be found by weighing an empty and dry measuring cylinder, filling it with glycerol and re-weighing to find the mass of glycerol. Then divide the mass of glycerol by the volume to find its density.

d Your graph of $v/\text{m s}^{-1}$ against r^2/m^2 should be a straight line passing through the origin and the point $(9.0 \times 10^{-6}\,\text{m}^2,\ 9.0 \times 10^{-2}\,\text{m s}^{-1})$.

e Gradient $= \dfrac{(9.0 - 0.0) \times 10^{-2}\,\text{m s}^{-1}}{(9.0 - 0.0) \times 10^{-6}\,\text{m}^2} = 1.0 \times 10^4\,\text{m}^{-1}\,\text{s}^{-1}$

$$\text{Viscosity } \eta = \frac{2(\rho_s - \rho_l)g}{9 \times \text{gradient}}$$

$$= \frac{2(7800 - 1260)\,\text{kg m}^{-3} \times 9.8\,\text{m s}^{-2}}{9 \times 1.0 \times 10^4\,\text{m}^{-1}\text{s}^{-1}}$$

$$= 1.4\,\text{kg m}^{-1}\,\text{s}^{-1}$$

$$= 1.4\,\text{Pa s}$$

Exam tip

It is often sensible to label the axes of a graph in the appropriate SI units to make calculating the gradient easier.

Knowledge check 65

Show that $\text{kg m}^{-1}\,\text{s}^{-1}$ is equivalent to Pa s.

f The following points should be made:
- As the student has not taken repeat readings for the diameters, it is not possible to determine what the uncertainty in these values is. It is likely to be greater than that of the instrument (0.01 mm), so let us assume that it is 0.02 mm. In that case the percentage uncertainty in the smallest diameter value is $\dfrac{0.02\,\text{mm}}{2.53\,\text{mm}} \times 100\% = 0.8\%$, giving a percentage uncertainty in r^2 of $2 \times 0.8\% = 1.6\%$.
- The smallest time value, as recorded, is 2.26 ± 0.05 s. In reality, we cannot consider the uncertainty to be less than human reaction time, which we can take as 0.1 s. The percentage uncertainty in the smallest time is therefore $\dfrac{0.1\,\text{s}}{2.26\,\text{s}} \times 100\% = 0.4\%$. As the '20 cm' for the distance was probably found from two readings using a mm scale, we can assume that the uncertainty is ±2 mm. This gives a percentage uncertainty of $\dfrac{2\,\text{mm}}{200\,\text{mm}} \times 100\% = 1.0\%$. The percentage uncertainty in the terminal velocity is therefore $0.4\% + 1.0\% = 1.4\%$.
- As we have had to make a number of assumptions, it would be reasonable to say that the uncertainties in the measurements of the diameter and the terminal velocity have more or less the same effect on the value for the viscosity.

g The viscosity of liquids such as glycerol decreases significantly with temperature, and so the temperature at which the viscosity was determined should always be stated.

Exam tip

Remember that when a quantity is squared its percentage uncertainty is doubled and that in a quotient the percentage uncertainty is the sum of the percentage uncertainties of the numerator and denominator.

Core practical 5

Determining the Young modulus of a material

Essential theory

The Young modulus E of a material is defined by the equation:

$$E = \frac{\text{stress}}{\text{strain}} = \frac{F/A}{\Delta l/l} = \frac{Fl}{A\Delta l}$$

Rearranging, we get:

$$F = \frac{EA}{l}\Delta l$$

A graph of the applied force F on the y-axis against the extension Δl on the x-axis will have a gradient EA/l. If we measure the length l, and determine A from a measurement of the diameter d, we can then determine E.

Planning

The extension of even a thin copper wire is only of the order of millimetres, which makes its measurement tricky. Although special apparatus, incorporating a vernier or micrometer, can be used, a reasonable value can be obtained for the Young modulus by the simple apparatus shown in Figure 44. This has the added advantage that the behaviour of the wire can be investigated up to when it breaks — but see 'Safety' below!

Figure 44

The following procedure should be carried out:

- The thin (e.g. 26 swg or 28 swg) copper wire should be tightly clamped to avoid slipping and should be as long as possible — usually about 3 m.
- Weights are added until the wire is just taut and the initial reading of the marker is taken.
- The extension can then be determined for additional loads.

Safety

The following precautions should be observed:

- Place a suitable 'crash pad' under the weights to prevent them hitting the floor directly when the wire breaks.
- Do not put your feet under the weights.
- Wear goggles for protection when the wire breaks.

Worked example

Table 11 shows a typical set of measurements. As the extensions are small, an attempt has been made to record the extensions to a precision of 0.5 mm.

Table 11

Mass added/kg	F/N	Δl/mm	Diameter d/mm
0.50	4.9	1.5	0.38, 0.37, 0.39, 0.38,
1.00	9.8	2.5	0.39, 0.37
1.50	14.7	3.0	Average d = 0.38 ±0.01 mm
2.00	19.6	4.0	
2.50	24.5	5.5	**Length l/m**
3.00	29.4	6.0	2.95, 2.93
3.50	34.3	7.5	Average l = 2.94 ± 0.01 m
4.00	39.2	10.0	
5.00	Large extension before wire breaks		

a Explain how the values for the diameter and the length of wire shown in the table could be determined.

b Plot a graph of load on the *y*-axis against extension on the *x*-axis, using error bars to reflect the precision of ±0.5 mm in the values of extension. (Although the extension is the dependent variable, it is conventional to plot load against extension or stress against strain.)

c Use the gradient of your graph and the data in the table to determine a value for the Young modulus for copper.

d Estimate the uncertainty in your value for the Young modulus.

Answer

a The length can be found using three metre rules (or a 3 m tape measure if available). Even if the precision is only 1 cm, this gives rise to an uncertainty of less than 1%. The length should be measured *from the point where the wire is clamped to the reference point on the marker*.

The diameter should be found in *different places and at different orientations* along the length of the wire using a digital vernier or micrometer having a precision of 0.01 mm or better.

b Your graph should look like Figure 45.

→

Exam tip

It is good practice to try to interpolate between the scale readings of an instrument if possible, e.g. to 0.5 mm on a millimetre scale or to 0.1°C on a thermometer having graduations of 1°C.

Knowledge check 67

Calculate the stress and strain in the wire when it is loaded with a mass of 2.00 kg. Use your values to calculate a value for the Young modulus.

Exam tip

Note that to get full marks the phrases in italics *must* be included.

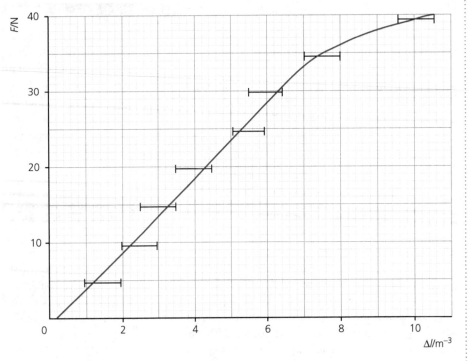

Figure 45

Knowledge check 68
Show that the units in the equation for E are equivalent to Pa.

c Determine the gradients for lines of maximum and minimum slope that fall within the error bars. Use the average of these values for your gradient value. This should be approximately $4800 \pm 400\,\text{N}\,\text{m}^{-1}$.

$$E = \text{gradient} \times \frac{l}{A} = 4800\text{N}\,\text{m}^{-1} \times \frac{2.94\,\text{m}}{\pi\left(\dfrac{0.38 \times 10^{-3}\,\text{m}}{2}\right)^2}$$

$$= 1.2(4) \times 10^{11}\,\text{Pa}$$

d % uc in diameter $= \dfrac{0.01\,\text{mm}}{0.38\,\text{mm}} \times 100\% = 2.6\%$

% uc in area $A = 2 \times 2.6\% = 5.2\%$

% uc in $l = \dfrac{0.01\,\text{m}}{2.94\,\text{m}} \times 100\% = 0.3\%$

% uc in gradient $= \dfrac{400\text{N}\,\text{m}^{-1}}{4800\text{N}\,\text{m}^{-1}} \times 100\% = 8.3\%$

% uc in $E = (5.2 + 0.3 + 8.3)\% = 13.8\%$

Uncertainty in $E = 13.8\%$ of $1.2(4) \times 10^{11}\,\text{Pa} = \pm 0.17 \times 10^{11}\,\text{Pa}$

This gives a value for E of $(1.24 \pm 0.17) \times 10^{11}\,\text{Pa}$ or, more realistically, $(1.2 \pm 0.2) \times 10^{11}\,\text{Pa}$.

Exam tip

Remember that, when a quantity is squared, its percentage uncertainty is doubled. In a product, the percentage uncertainty is the sum of the percentage uncertainties of the individual terms. And in a quotient, the percentage uncertainty is the sum of the percentage uncertainties of the numerator and denominator.

Core practical 6

Determining the speed of sound in air using a two-beam oscilloscope, signal generator, speaker and microphone

Essential theory

The wavelength λ, frequency f and speed c of a wave are related by the equation:

$$c = f\lambda$$

If we can measure the wavelength of sound waves for different frequencies, we can use the equation to find the speed of sound.

Planning

The apparatus is arranged as shown in Figure 46.

Signal generator

Microphone

Speaker

Rule secured to bench with modelling clay

Two-beam oscilloscope

Figure 46

The following procedure should be carried out:

- Connect the signal generator to one input of the dual-beam oscilloscope and the loudspeaker to the other input.
- Adjust the time base and voltage sensitivity to give about two complete waves of reasonable height on the screen.
- Move the microphone away from the speaker until the wave traces are in exact antiphase. Note the position of the microphone on the scale.
- Gradually move the microphone further away until the wave traces are in exact antiphase again. The distance that the microphone has moved is then one wavelength.
- Check the frequency with a frequency meter or by using the oscilloscope.
- Repeat the experiment at different frequencies.

Safety

The following precautions should be observed:

■ Even though the output from the signal generator is low-voltage, care should be taken to avoid any short circuits.

■ Mains equipment such as the signal generator and oscilloscope should always be treated with respect.

Knowledge check 70

Figure 47 shows a trace on an oscilloscope screen. Determine the frequency and potential difference of the signal being displayed.

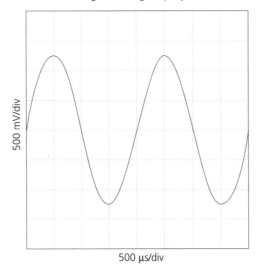

500 μs/div

Figure 47

Exam tip

It is important to record the temperature, because the speed of sound increases with the temperature of the air.

Worked example

Table 12 shows a typical set of measurements taken at 21°C.

Table 12

Frequency f/Hz	(1/f)/ms	Wavelength λ/m	Average λ/m
500		0.663, 0.673	0.668
600		0.561, 0.553	0.557
700		0.483, 0.475	0.479
800		0.417, 0.425	0.421
1000		0.331, 0.341	0.336
1200		0.277, 0.273	0.275

a Draw two sketches of an oscilloscope screen displaying two waves that are (i) in phase and (ii) in antiphase. Each sketch should show two complete waves.

b Explain how a graph of λ on the y-axis against 1/f on the x-axis can be used to determine the speed of sound. →

c Add values for $1/f$ to the table and plot a graph of λ against $1/f$. Use your graph to determine the speed of sound.

d The frequency was measured with a frequency meter, which the manufacturer states has a tolerance of ±10 Hz. By considering the measurements at 800 Hz, estimate the uncertainty in your value for the speed of sound.

e Explain how you would adapt the experiment if frequencies in the range 2 kHz to 5 kHz were to be used.

Answer

a Your sketch should look like Figure 48.

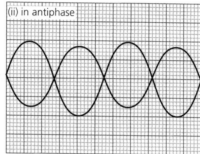

Figure 48

b Rearranging the equation $c = f\lambda$ gives us:

$$\lambda = c \times \frac{1}{f}$$

Therefore, a graph of λ against $1/f$ will be a straight line through the origin of gradient c.

c The $1/f$ values, in ms, are 2.00, 1.67, 1.43, 1.25, 1.00 and 0.83. Your graph should be a straight line passing through the origin and the point (1.75 ms, 0.590 m). Using these points for the gradient gives us:

$$c = \text{gradient} = \frac{(0.590 - 0.000)\,\text{m}}{(1.75 - 0.00) \times 10^{-3}\,\text{s}} = 337\,\text{m s}^{-1}$$

d The percentage uncertainty in 800 Hz is $\frac{10\,\text{Hz}}{800\,\text{Hz}} \times 100\% = 1.25\%$

The uncertainty in the corresponding wavelength is ½ × range,
i.e. ½ × 0.008 m = 0.004 m

This gives a percentage uncertainty for the wavelength of $\frac{0.004\,\text{m}}{0.421\,\text{m}} \times 100\% = 0.95\%$

The percentage uncertainty in the speed is therefore 1.25% + 0.95% = 2.2%

The uncertainty in the speed is then 2.2% × 337 m s^{-1} = ±7 m s^{-1}

➜

Exam tip

Remember that, for a quotient, the percentage uncertainty is the sum of the percentage uncertainties of the numerator and denominator.

We should therefore give the speed as $(337 \pm 7)\,\text{m}\,\text{s}^{-1}$.

(In practice, plotting the graph averages six readings, and so the uncertainty is probably a bit less than this.)

e A frequency of, say, 3 kHz would give a wavelength of the order of 10 cm. It would therefore be better to move the microphone back through several wavelengths so that a larger distance was measured, thus reducing the percentage uncertainty.

Exam tip

Always try to make measured quantities as large as possible so that the percentage uncertainty in the measurement is minimised.

Core practical 7

Investigating the effects of length, tension and mass per unit length on the frequency of a vibrating string or wire

Essential theory

If the wire in Figure 49 is plucked, it will vibrate at its fundamental frequency, with nodes at each end and one antinode at its centre. The distance l between the nodes is $\frac{1}{2}\lambda$ and so $\lambda = 2l$.

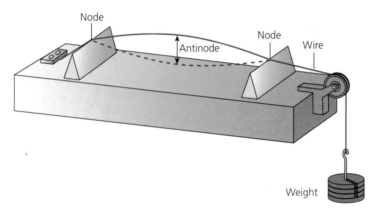

Figure 49

For a wire or string of mass per unit length μ under a tension T, the fundamental frequency f is given by the equation:

$$f = \frac{1}{2l}\sqrt{\frac{T}{\mu}}$$

We can see that f depends on *three* variables: l, T and μ. We must therefore keep two of these variables constant while the relationship between f and the third variable is investigated.

Knowledge check 71

Show that the fundamental frequency of a wire having a length of 50 cm and a mass per unit length of $2\,\text{g}\,\text{m}^{-1}$ under a tension of 10 N is about 70 Hz.

Planning

For a given wire, μ will be constant. Therefore, we can either keep l constant and investigate how f depends on T, or keep T constant to see how f depends on l. Two ways of setting the wire vibrating are shown in Figure 50.

Figure 50

Exam tip

It is good practice to determine the frequency by increasing it until a large vibration is observed and recording this value. Then increase the frequency a bit more and gradually reduce it again until a large vibration is once more observed, giving a second reading. The two readings can then be averaged.

- If the *length is kept constant*, the frequency can be adjusted so that the wire vibrates at its fundamental frequency for a series of weights added to the wire.
- Likewise, if the wire is under the *tension of a fixed load*, the fundamental frequency can be found for different lengths of the wire.
- Appropriate graphs can then be plotted to test the relationship.

Safety

Some basic precautions should be observed — see part **a** of the worked example below.

Worked example

Table 13 shows a typical set of measurements for a steel wire of length $l = 0.500\,\text{m}$.

Table 13

Mass/kg	T/N	Frequency/Hz	Average f/Hz	$f^2/10^3\,\text{Hz}^2$
0.50	4.9	46, 50	48	
1.00	9.8	69, 63	66	
1.50	14.7	85, 77	81	
2.00	19.6	90, 98	94	
3.00	29.4	111, 117	114	
4.00	39.2	137, 129	133	

a State any safety precautions you would adopt when doing this experiment.

b Explain how a graph of f^2 on the y-axis against T on the x-axis can be used to show the relationship between T and f.

c Add values for f^2 to the table and plot a graph of f^2 against T. What does your graph enable you to deduce about the relationship between T and f?

d Use the gradient of your graph to show that the mass per unit length μ of the wire is about $2\,\text{g}\,\text{m}^{-1}$.

e Describe how you would check your value for μ by direct measurement.

f Sketch the graph you would plot to investigate the relationship between f and l for a fixed load.

g Explain briefly how you would investigate the relationship between f and μ. You should include a sketch of the graph you would plot.

→

Content Guidance

Exam tip

You must state that the graph *passes through the origin* when testing for a proportional relationship. No marks will be awarded for just saying 'a straight line'.

Answer

a The following precautions should be observed:
  - Place a suitable 'crash pad' under the weights to prevent them hitting the floor directly if the wire breaks.
  - Do not put your feet under the weights (for the same reason).
  - Wear goggles for protection in case the wire breaks.

b Squaring the formula $f = \dfrac{1}{2l}\sqrt{\dfrac{T}{\mu}}$ on both sides gives us $f^2 = \dfrac{T}{4l^2\mu}$. Therefore, a graph of f^2 against T should be a straight line through the origin with a gradient $\dfrac{1}{4l^2\mu}$.

c The values for $f^2/10^3\,\text{Hz}^2$ are: 2.3, 4.4, 6.6, 8.8, 13.0 and 17.7. Your graph of f^2 against T should be a straight line *through the origin* and the point (40 N, $18 \times 10^3\,\text{Hz}^2$). This shows that $f^2 \propto T$ or $f \propto \sqrt{T}$ as in the equation.

d From answer **b** the gradient is $\dfrac{1}{4l^2\mu}$.

From answer **c** its value is $\dfrac{(18.0 - 0.0) \times 10^3\,\text{s}^{-2}}{(40.0 - 0.0)\,\text{N}} = 450\,\text{N}^{-1}\,\text{s}^{-2}$.

$$\mu = \frac{1}{\text{gradient} \times 4l^2} = \frac{1}{450\,\text{N}^{-1}\,\text{s}^{-2} \times 4 \times 0.250\,\text{m}^2}$$

$$= 2.2 \times 10^{-3}\,\text{kg}\,\text{m}^{-1} = 2.2\,\text{g}\,\text{m}^{-1}$$

e As μ is only $2\,\text{g}\,\text{m}^{-1}$, the mass of, say, 5 m of wire should be found ($\approx 10\,\text{g}$) using a balance of sensitivity 0.1 g or better.

f For constant T, we need to plot f against $1/l$ to test whether $f \propto \dfrac{1}{l}$. This is shown in Figure 51.

g To investigate how f is related to μ, the fundamental frequency should be found for different wires (i.e. different μ), keeping l and T the same for each wire. A graph of f^2 against $1/\mu$ should then be plotted, as in Figure 52.

Knowledge check 72

Show that the units of $1/(\text{N}^{-1}\,\text{s}^{-2} \times \text{m}^2)$ are equivalent to $\text{kg}\,\text{m}^{-1}$.

Figure 51

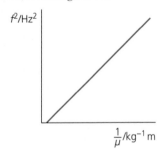

Figure 52

Core practical 8

Determining the wavelength of light from a laser or other light source using a diffraction grating

Essential theory

When monochromatic light, such as a laser beam, is shone through a diffraction grating, a pattern of lines is seen, as shown in Figure 53(a). There is a central bright maximum, with further maxima of decreasing intensity on either side. This is represented diagrammatically by Figure 53(b).

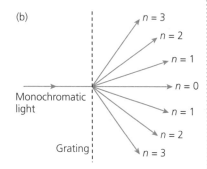

Figure 53

The angle θ of a particular maximum from the central maximum is given by the equation:

$$n\lambda = d \sin \theta$$

where n is the 'order' of the maximum (i.e. $n = 1$ is the first maximum on each side of the central maximum, and so on), d is the separation of the lines of the grating and λ is the wavelength of the light.

Planning

The experimental arrangement is shown in Figure 54. The light source is a laser emitting green light.

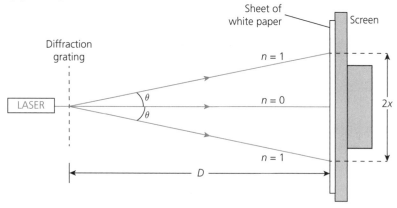

Figure 54

Knowledge check 73

Show that the angle of diffraction of the first order for light of wavelength 550 nm transmitted through a grating of spacing 300 lines per mm is about 10°.

You should proceed as follows:

- Adjust the distance D between the grating and the screen until a number of orders can be seen, sufficiently far apart for their separation to be measured with a millimetre scale.
- Measurements should *not* be attempted directly — the centres of the maxima should be marked on a piece of paper secured to the front of the screen and then processed after the laser has been switched off (see 'Safety' below).
- Measure the distance apart, $2x$, between the maxima of each order, and then determine θ for each order using $\tan\theta = x/D$.

Exam tip

You should always try to make measurements that are as large as possible to minimise experimental uncertainty.

Safety

Lasers are potentially **dangerous**! Strict precautions should be observed in their use:

- Never look directly into the beam, or allow anyone else to.
- Never point the beam at anyone.
- Avoid reflections of the beam.
- Always keep your back to the laser when it is on.
- Always switch off the laser when not in use.
- Always work in a well-lit room to prevent the pupils of your eyes dilating.

Worked example

The photograph in Figure 53(a) is *one-quarter* full scale.

a The distance from the grating to the screen was 200 ± 2 mm. Take such measurements as are necessary to determine the angles of diffraction for the first three orders of the green light. Record your measurements in a suitable table.

b The manufacturer states that the grating has a spacing of 300 ± 5 lines per mm. What value does this give for the line separation d?

c Use this, with your values for θ, to find an average value for the wavelength of the green light.

d Estimate the uncertainty in your value for the wavelength.

e Discuss two ways in which you could reduce this uncertainty.

Exam tip

You must be able both to draw, and to use, scaled diagrams.

Answer

a Your table should look like this:

Order	2x/mm in photo	2x/mm actual	x/mm	tan θ	θ/°
1	17.0	17.0 × 4 = 68.0	34.0	0.170	9.7
2	34.5	34.5 × 4 = 138.0	69.0	0.345	19.0
3	53.0	53.0 × 4 = 212.0	106.0	0.530	27.9

b Line separation $d = 1/(300 \times 10^3)\,\text{m} = 3.33 \times 10^{-6}\,\text{m}$

Percentage uncertainty in manufacturer's value is $(5/300) \times 100\% = 1.7\%$

Uncertainty in $d = 1.7\%$ of $3.33 \times 10^{-6}\,\text{m} = \pm0.07 \times 10^{-6}\,\text{m}$

The value for d should therefore be stated as $(3.33 \pm 0.07) \times 10^{-6}\,\text{m}$

→

Exam tip

Try to interpolate between scale divisions, e.g. to 0.5 mm on a millimetre scale, as here.

c From $n\lambda = d \sin\theta$ we obtain $\lambda = \dfrac{d\sin\theta}{n}$

  $n = 1$: $\lambda = \dfrac{3.33\times10^{-6}\,\text{m} \times \sin 9.7°}{1} = 564\,\text{nm}$

  $n = 2$: $\lambda = \dfrac{3.33\times10^{-6}\,\text{m} \times \sin 19.0°}{2} = 544\,\text{nm}$

  $n = 3$: $\lambda = \dfrac{3.33 \times 10^{-6}\,\text{m} \times \sin 27.9°}{3} = 520\,\text{nm}$

  Average $\lambda = 543$ nm

d The range of values is $(564 - 520)$ nm, so:

  random uncertainty (from measurement of x) = ½ range = ±22 nm

  % random uc = $(22\,\text{nm}/543\,\text{nm}) \times 100\% = 4.0\%$

  % systematic uc in d = 1.7% (see **b** above)

  % systematic uc in screen distance = $(2\,\text{mm}/200\,\text{mm}) \times 100\% = 1.0\%$

  % uc in λ = 4.0% + 1.7% + 1.0% = 6.7%

  Uncertainty in λ = 6.7% × 543 nm = 36 nm

  We can therefore say that $\lambda = 543 \pm 36$ nm

e The uncertainty could be reduced by:

  ▪ moving the screen further away (provided the pattern remains visible) so that the distance between the maxima is greater

  ▪ using a grating with smaller line separation, e.g. 600 lines per mm, so that the maxima are further apart

Core practical 9

Investigating the relationship between the force exerted on an object and its change of momentum

Essential theory

Newton's second law states that the resultant force acting on an object is equal to the rate of change of momentum of the object. Mathematically, this can be represented by:

  $F = \dfrac{\text{d}(mv)}{\text{d}t}$ or $F = m\dfrac{\Delta v}{\Delta t}$ for a fixed mass

If a graph of F on the y-axis is plotted against $\dfrac{\Delta v}{\Delta t}$ on the x-axis, a straight line through the origin of gradient m will then confirm the relationship between force and change of momentum.

Planning

One experimental arrangement is shown in Figure 55.

Knowledge check 74

Assume that the diffraction grating and screen distance in the worked example are kept the same.

a Explain the difference you would observe in the distances between the maxima if the green laser was replaced by a red laser emitting a wavelength of 693 nm.

b (A-level only) Determine how many orders can be observed with this red laser.

Exam tip

For a question like **e**, which asks you to *discuss*, you must state what you would do *and* explain how this would reduce the uncertainty.

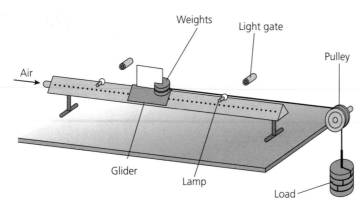

Figure 55

Exam tip

You should make sure you are familiar with the use of light gates, a data logger (or computer interface) to collect data and computer software to analyse the data.

Points to remember are as follows:

- Light gates are used to measure the times t_1 and t_2 for the card of length l to pass through them.
- The velocity of the air-track glider can then be calculated using $v = l/t$.
- If the light gates are connected to a data logger, a print-out like that shown in Figure 56 can be obtained, from which the time Δt between gates can be found.

Figure 56

Safety

- A shock pad should be placed under the weights to prevent any damage in case they hit the floor.
- Do not put your feet under the weights.

Knowledge check 75

In Figure 55, each of the five weights and the hanger have a mass of 10 g and the air-track glider and card have a mass of 440 g. Show that the glider will have an acceleration of about $0.8\,\mathrm{m\,s^{-2}}$.

Worked example

In Figure 55, the card has a length of 150 mm and the mass of the glider and card is 440 g. Each of the five weights and the hanger has a mass of 10 g. Table 14 shows a typical set of results.

Table 14

F/N	t_1/s	t_2/s	v_1/m s^{-1}	v_2/m s^{-1}	Δt/s	$(\Delta v/\Delta t)$/m s^{-2}
0.10	0.38	0.27	0.39	0.56	0.85	0.20
0.20	0.28	0.20			0.52	
0.29	0.22	0.16			0.45	
0.39	0.19	0.14			0.36	
0.49	0.16	0.12			0.32	
0.59	0.14	0.11			0.24	

a Explain how you would apply the forces shown and, at the same time, keep the mass being accelerated constant.

b Complete the table. You could set up a spreadsheet to 'crunch' the numbers.

c Plot a graph of F on the y-axis against $(\Delta v/\Delta t)$ on the x-axis.

d Discuss the extent to which your graph confirms that the rate of change of velocity is proportional to the force.

e Determine the gradient of your graph. What physical quantity does it represent?

f Explain how your answers to **d** and **e** show that the force is equal to the rate of change of momentum.

g Suggest why the last two values show a little scatter from the line of best fit.

Answer

a The secret is to move the weights, in turn, from the glider to the weight hanger.

- The total mass is the mass of the glider and card (440 g) plus the mass of five weights and hanger (60 g), i.e. 500 g.
- With just the hanger pulling the glider and five weights, the force is $0.01 \text{ kg} \times 9.8 \text{ N kg}^{-1} = 0.10 \text{ N}$. This force is accelerating a total mass of 500 g.
- If one mass from the glider is then transferred to the hanger, the force is $0.02 \text{ kg} \times 9.8 \text{ N kg}^{-1} = 0.20 \text{ N}$, but the total mass being accelerated is still 500 g. And so on.

b The completed table is shown below.

F/N	t_1/s	t_2/s	v_1/m s^{-1}	v_2/m s^{-1}	Δt/s	$(\Delta v/\Delta t)$/m s^{-2}
0.10	0.38	0.27	0.39	0.56	0.85	0.20
0.20	0.28	0.20	0.54	0.75	0.52	0.40
0.29	0.22	0.16	0.68	0.94	0.45	0.58
0.39	0.19	0.14	0.79	1.07	0.36	0.78
0.49	0.16	0.12	0.94	1.25	0.32	0.97
0.59	0.14	0.11	1.07	1.36	0.24	1.21

c You should get a straight-line graph through the origin and the point (1.2 m s^{-1}, 0.60 N).

d As the graph is a straight line *through the origin* it shows that rate of change of velocity is proportional to the force.

e Drawing a triangle between the origin and the point (1.2 m s^{-1}, 0.60 N) gives:

$$\text{gradient} = \frac{(0.60 - 0.00) \text{ N}}{(1.20 - 0.00) \text{ m s}^{-1}} = 0.500 \text{ kg} = 500 \text{ g}$$

This represents the total mass being accelerated.

f Answer **d** shows $F \propto \Delta v/\Delta t$ and answer **e** shows that the constant of proportionality is M. Hence $F = M(\Delta v/\Delta t) = (\Delta[Mv])/\Delta t$ = rate of change of momentum.

g The scatter of the last two points from the line of best fit is probably due to t_1 and t_2 having small values, leading to less accurate values for v_1 and v_2.

Exam tip

Note that in part **d** there are two important points to be made:

- The mass is constant.
- The line goes through the origin.

Marks will be lost if these points are not included.

Exam tip

Remember to draw a large triangle to find a gradient.

Core practical 10

Using ICT to analyse collisions between small spheres, e.g. ball bearings on a table top

The Edexcel specification prescribes that you should develop practical skills to enable you to:

- use appropriate software and tools to process data, carry out research and report findings
- use online and offline research skills, including websites, textbooks and other printed scientific sources of information
- correctly cite sources of information

The purpose of this core practical is to provide you with the opportunity to do some research of your own. How you approach this will very much depend on the apparatus available to you and your own ICT ability.

Essential theory

You should be familiar with the conservation of linear motion and its application to collisions in two dimensions, such as that between snooker balls, as shown in Figure 57. You should also be able to analyse the motion of the balls.

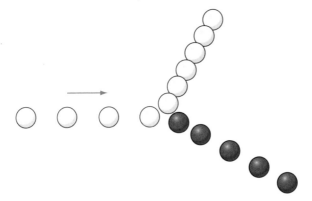

Figure 57

Planning

There are a number of ways that you can perform this practical exercise. As suggested in the specification, you could use ball bearings on a table top. The procedure might be to:

- use a ramp to 'launch' a ball bearing at a stationary ball bearing
- film the collision with, say, a mobile phone
- analyse the motion with suitable software
- use the analysis to investigate whether linear momentum has been conserved

The analysis could be undertaken using the free software *Tracker: Video Analysis and Modeling Tool* available on http://physlets.org/tracker/. There are several examples on YouTube that you might like to research.

Safety

The following precaution should be observed:

■ Ball bearings falling on to the floor could be a hazard, and so it is recommended that the experiment is conducted in a large tray.

Worked example

Figure 57 shows a collision between two snooker balls having the same nominal mass (142 ± 3 g) taken from a stroboscopic photograph. The red ball is initially stationary. The stroboscope flashes at a rate of 50 Hz and the diagram is *one-tenth* scale.

a Show that the speed of the white ball before collision is approximately 5.7 m s^{-1}. Explain how you did this.

b Estimate the percentage uncertainty in your value.

c Determine the speed of each ball after the collision.

d Calculate the magnitude of the momentum of each ball before and after the collision.

e Measure the angles between the initial direction of the white ball and each of the two balls after the collision.

f Hence investigate the extent to which linear momentum has been conserved in the collision.

Answer

a The distance for four flashes in the diagram is 34 mm, so the actual distance is 340 mm.

Time between flashes = 1/(50 Hz) = 0.020 s.

Time between four flashes = 3 × 0.020 s = 0.060 s

Speed = 0.34 m/0.060 s = 5.67 m s^{-1} ≈ 5.7 m s^{-1}

b The uncertainty in 34 mm is ±1 mm.

% uc in speed = % uc in distance = (1 mm/34 mm) × 100% = 3%

c After collision:

speed of white ball = (10 × 15 mm)/(3 × 0.020 s) = 2.5 m s^{-1}

speed of red ball = (10 × 27 mm)/(3 × 0.020 s) = 4.5 m s^{-1}

d Before collision:

$p_w = 0.142$ kg × 5.7 m s^{-1} = 0.81 kg m s^{-1} and

$p_r = 0$

After collision:

$p_w = 0.142$ kg × 2.5 m s^{-1} = 0.36 kg m s^{-1} and

$p_r = 0.142$ kg × 4.5 m s^{-1} = 0.64 kg m s^{-1}

e $\theta_w = 60°$ and $\theta_r = 30°$

→

f Resolving momentum normal to initial direction after collision:

resultant $= (0.36\,\mathrm{kg\,m\,s^{-1}})\sin 60°$ ('up') $- (0.64\,\mathrm{kg\,m\,s^{-1}})\sin 30°$ ('down')

$= (0.31 - 0.32)\,\mathrm{kg\,m\,s^{-1}} \approx 0 =$ component before collision

Resolving momentum along initial direction after collision:

resultant $= (0.36\,\mathrm{kg\,m\,s^{-1}})\cos 60° + (0.64\,\mathrm{kg\,m\,s^{-1}})\cos 30°$

$= (0.18 + 0.55)\,\mathrm{kg\,m\,s^{-1}} = 0.73\,\mathrm{kg\,m\,s^{-1}}$

The latter differs from the initial momentum of $0.81\,\mathrm{kg\,m\,s^{-1}}$ by about 10%. The uncertainties in the measured distances are about 3%, 7% and 4%, respectively, and the masses could differ by as much as $(6\,\mathrm{g}/142\,\mathrm{g}) \times 100\% = 4\%$. There is also an uncertainty in the measurement of the angles. A difference of 10% in the momentum could therefore be easily accounted for by experimental uncertainty.

The purpose of this core practical is that the analysis should be done using ICT. However, in the written examination, a question like this could be set, which tests your understanding of the principles behind the experiment and your ability to draw a suitable conclusion from your results.

Exam tip

When evaluating the results of an experiment, as much quantitative evidence as possible should be used to substantiate your argument.

Core practical 11

Using an oscilloscope or data logger to display and analyse the potential difference (p.d.) across a capacitor as it charges and discharges through a resistor

Essential theory

You should be familiar with the formula for a capacitor of capacitance C being discharged through a resistor of resistance R. The potential difference (p.d.) V across the capacitor decreases exponentially with time t according to the equation:

$V = V_0 e^{-t/RC}$

where V_0 is the p.d. when $t = 0$. The product RC is called the time constant for the discharge. This is the time for the p.d. to fall to $1/e$ of its initial value, i.e. $0.368V_0$.

You should also be aware that the p.d. across a capacitor being charged through a resistor increases exponentially with time, although you do not have to know the formula for this. The p.d. rises to 0.632 of its maximum value in one time constant.

Planning

As capacitors normally have values of the order of microfarads (µF), the time constant will be short unless the resistance is very large — see knowledge check 77. This means that sometimes it is not possible to investigate the discharge by collecting the data as you go, and so we have to use an oscilloscope, or data logger and computer, to show the charge and discharge.

Knowledge check 77

a Show that the time constant for a 220 µF capacitor discharging through a 3.3 kΩ resistor is of the order of 1 s.

b Use the exponential equation to explain why RC must have the unit of seconds.

c Confirm this by considering the units of resistance and capacitance.

Exam tip

Remember that the product RC gives an order of magnitude for the time scale for the discharge of a capacitor.

Figure 58

The experimental arrangement is shown in Figure 58.

■ The capacitor is charged through the resistor by connecting the switch to contact A and then discharged through the same resistor by moving the switch to contact B.

■ The p.d. across the capacitor is measured by the voltage sensor and recorded by the data logger.

■ The data can then be analysed by the computer.

Safety

■ **Danger!** If an electrolytic capacitor is being used, its polarity *must* be observed, i.e. the end marked positive (+) must be connected to the positive side of the power supply.

■ Although the p.d. is low voltage, care should be taken to avoid any short circuits.

■ Check that the value of the p.d. is compatible with the voltage sensor.

Worked example

Figure 59 shows a printout of the charge and discharge of a $22\,\mu F$ capacitor through a $3.3\,k\Omega$ resistor.

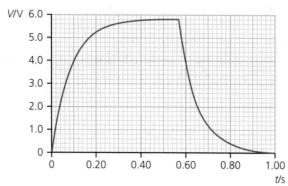

Figure 59

a The value of the resistor is checked with a digital ohmmeter and found to be $3.34\,k\Omega$. Use the discharge curve to determine an experimental value for the capacitance of the capacitor. Estimate the percentage uncertainty in your value.

b The manufacturer's tolerance is ±2% for the resistor and ±10% for the capacitor. Discuss the extent to which your values for the resistance and capacitance are compatible with these data.

Exam tip

Remember that the values stamped on resistors and capacitors are what are called 'nominal values' and are subject to a manufacturing tolerance. This is usually 1% or 2% for resistors, and can be as much as 10% or even 20% for capacitors.

Content Guidance

c Suggest how you could display the data to improve the accuracy of your value for C.

d Explain why processing the data in the form of a graph of $\ln(V/V)$ against t/s would enable you to determine a more accurate value for C.

e Discuss the advantages of using a data logger and computer rather than an oscilloscope for this experiment.

Answer

a You need to make the following points:
- The p.d. will fall from $5.8\,V$ to $(0.368 \times 5.8)\,V = 2.1\,V$ in a time RC.
- The capacitor starts to discharge at $0.57\,s$ and the p.d. drops to $2.1\,V$ after $0.65\,s$.
- This gives a time constant of $(0.65 - 0.57)\,s = 0.08\,s$.
- From time constant $= RC = 0.08\,s \Rightarrow C = 0.08\,s/(3.34 \times 10^3\,\Omega) = 2.4 \times 10^{-5}\,F = 24\,\mu F$.
- The small scale suggests it would be reasonable to take the uncertainty in RC as $0.01\,s$. This gives a percentage uncertainty of $(0.01\,s/0.08\,s) \times 100\% = 12.5\%$.
- Assuming the ohmmeter has an uncertainty of $0.01\,k\Omega$, the percentage uncertainty in R will be $(0.01\,k\Omega/3.34\,k\Omega) \times 100\% = 0.3\%$.
- This gives a percentage uncertainty in C of $(12.5 + 0.3)\% = 12.8\% = 13\%$.

(You may have been taught that the p.d. falls to $V_0/2$ in a time of $0.693RC$. This 'half-life' method is equally acceptable and should yield a similar value for C, bearing in mind the uncertainty in reading off the times.)

b The difference between the stated and measured values for R is:

$(3.34 - 3.3)\,k\Omega = 0.04\,k\Omega$

The percentage difference is therefore $(0.04\,k\Omega/3.3\,k\Omega) \times 100\% = 1.2\%$, which is well within the 2% tolerance stated.

The difference between the stated and measured values for C is $(24 - 22)\,\mu F$. This is a percentage difference of $(2\,\mu F/22\,\mu F) \times 100\% = 9\%$, which is within the 13% uncertainty calculated for C and within the stated tolerance of 10%.

c The data could be displayed on a larger scale by 'zooming in' on the points where the discharge starts and where the p.d. is $2.1\,V$ to get more accurate values for the times.

d $V = V_0 e^{-t/RC} \Rightarrow \ln V = \ln V_0 - \dfrac{t}{RC}$

A graph of $\ln(V/V)$ against t/s would give a straight line of gradient $-(1/RC)$, from which RC can be found. A more accurate value for C should be obtained, as a straight-line graph effectively averages several readings.

e An oscilloscope would not be able to:
- give a printout of the actual data
- process the data
- give data directly to enable a graph of $\ln(V/V)$ against t to be plotted

Exam tip

You must be able to convert a logarithmic or exponential equation into a linear form by taking natural logarithms on both sides of the equation.

Knowledge check 78

What is the current in the capacitor when it just starts to discharge?

Core practical 12

Calibrating a thermistor in a potential divider circuit as a thermostat

Essential theory

You should be familiar with the potential divider circuit. Figure 60 shows the circuit for a thermistor as resistor R_1. You should also know that the resistance of a negative temperature coefficient thermistor decreases with increasing temperature.

In this circuit:

$$V_{out} = V_{in} \times \frac{R_2}{R_1 + R_2}$$

If the resistance R_1 of the thermistor gets less as the thermistor warms up, the value of $R_1 + R_2$ will decrease, and so the output voltage V_{out} will increase. This increase in output voltage can be used trigger a switch, and so the circuit can be used as a thermostat.

Planning

The experimental arrangement is shown in Figure 61.

Figure 61

These are the points to remember:

- The value of the resistor should be chosen such that $R_2 \approx R_1$ (see knowledge check 80).
- The p.d. across the resistor is measured at 10°C intervals from about 10°C to 60°C.

Knowledge check 79

What is meant by a negative temperature coefficient thermistor? Sketch how its resistance varies with temperature.

Figure 60

Knowledge check 80

In Figure 60 V_{in} = 6.0 V and the resistor R_2 has a value of 330 Ω. The manufacturer states that the thermistor R_1 has a resistance of 470 Ω at a temperature of 25°C, with a tolerance of ±2%. Show that the output voltage is about 2.5 V at this temperature.

- The resistance of the thermistor at each temperature can then be calculated using the potential divider formula.
- A graph of output voltage against temperature should be drawn.
- The resistance of the thermistor at any desired temperature within the range investigated can then be found by reading off the value of V_{out} from the graph and substituting it into the equation to find R_1.

Safety

- Keep the wires well away from the heat source to avoid melting the insulation.
- Make sure the beaker is firmly supported so that it does not topple.
- Use a heat-resistant mat under the heat source.

Worked example

Table 15 shows a typical set of readings.

Table 15

$\theta/°C$	10	20	30	40	50	60
V_{out}/V	1.78	2.25	2.70	3.10	3.44	3.76

a Describe the technique you would use to make the readings as accurate as possible, and how you would get readings below room temperature.

b Plot a graph of V_{out} on the y-axis against θ on the x-axis.

c To what extent is the manufacturer's specification (see knowledge check 80) confirmed by the experimental data?

d The output switches an electronic circuit when it reaches 3.00 V. Show that a value of R_2 of about 270 Ω would be needed if the switch is to be used as a thermostat to switch off an immersion heater when the water temperature reaches 45°C.

e The nearest available preferred value resistor to 270 Ω is 220 Ω. Explain how a switch-off at precisely 45°C could be achieved.

Answer

a To try to get accurate readings:
  - stir the water continuously
  - as you approach the required temperature, turn the heat down and let the temperature stabilise
  - keep the thermistor away from the sides and bottom of the beaker

  Temperatures below room temperature can be achieved by adding small pieces of ice to the water.

b You should find that your plotted points lie on a gentle curve. You should draw a smooth curve of best fit through them.

Exam tip

Not all graphs are straight lines! 'Forcing' a straight line through these points would lose marks. You should practice drawing *smooth* curves.

Exam tip

These are standard techniques applicable to all experiments involving heating a liquid and recording its temperature.

c At 25°C the value of V_{out} = 2.48 ± 0.02 V.

$$V_{out} = V_{in} \times \frac{R_2}{R_1 + R_2} \Rightarrow R_1 = \left(\frac{V_{in} \times R_2}{V_{out}}\right) - R_2$$

so:

$$R_1 = \left(\frac{6\,V \times 330\,\Omega}{2.48\,V}\right) - 330\,\Omega = 798\,\Omega - 330\,\Omega = 468\,\Omega$$

If the tolerance in the manufacturer's value of 470 Ω is ±2%, the value could be between 461 Ω and 479 Ω. The experimental value lies within this range, and therefore confirms the manufacturer's specification.

d At 45°C, V_{out} = 3.28 V so:

$$R_1 = \left(\frac{6\,V \times 330\,\Omega}{3.28\,V}\right) - 330\,\Omega = 604\,\Omega - 330\,\Omega = 274\,\Omega$$

To give an output voltage of 3.00 V, which is exactly half the input voltage of 6.0 V, the value of the resistor must equal the resistance of the thermistor, i.e. 274 Ω.

e This could be achieved by using a 220 Ω resistor and adding a small potentiometer (rheostat) in series with it. The thermistor should be immersed in water at 45°C and the potentiometer adjusted until the switch is just activated. The 'thermostat' has then been calibrated to switch off at 45°C.

Core practical 13

Determining the specific latent heat of a phase change

Essential theory

You should be familiar with the concept of a substance requiring energy in order to change phase (i.e. from solid to liquid, or from liquid to vapour). The amount of energy ΔE required depends on the mass Δm and a property of the substance, called its specific latent heat of fusion L_f for melting and specific latent heat of vaporisation L_v for changing into a vapour. We then have (using L to stand for either L_f or L_v):

$$\Delta E = L\Delta m$$

From this equation, we can see that the unit for L is $J\,kg^{-1}$.

There are numerous ways of determining L experimentally, either by what is called the 'method of mixtures' or electrically. Both methods involve some fundamental principles of experimental physics, and so we will look at an example of each.

Knowledge check 81

How much energy is needed to melt a 10 g ice cube at 0°C if the specific latent heat of fusion for water is 330 kJ kg^{-1}?

Experiment A: Method of mixtures for latent heat of fusion

Planning

This method is a simple, but reasonably accurate, way of finding L_f for ice. The crucial precaution is to use small pieces of dry, crushed ice that is on the point of melting. The procedure is as follows.

- Warm some tap water until it is about 10°C above room temperature and then transfer 200 cm^3 (200 g) into an expanded polystyrene cup.
- Weigh the cup and water.
- Record the initial temperature θ_i of this water when it is about 5°C above room temperature.
- Quickly dry some crushed ice and add small pieces steadily to the water, stirring continuously, until the final temperature θ_f is about 5°C below room temperature.
- Re-weigh the cup and water to find the mass Δm of ice that has been added.

The latent heat of fusion can then be determined by equating the energy gained by the ice to the energy lost by the water in the cup (an application of the principle of conservation of energy).

Safety

- Take care not to topple the beaker when heating the water.
- Ensure that the cup does not topple (e.g. do not leave the thermometer resting inside it!).
- Use paper towels to mop up any spillages immediately (slip hazard).

Worked example

The following data were recorded:

- Mass of water in cup: $m = 200$ g
- Initial temperature of water in cup: $\theta_i = 24$°C
- Final temperature of water in cup: $\theta_f = 14$°C
- Mass of ice added: $\Delta m = 22.6$ g
- Specific heat capacity of water: $c = 4200 \, \text{J kg}^{-1} \text{K}^{-1}$

a Explain two advantages of using an expanded polystyrene cup in this experiment.

b Discuss how the effect of energy lost to, or gained from, the surroundings is minimised in this experiment.

c Explain why is it essential to add *small* pieces of ice to the cup of the water.

d Show that:
- the energy lost by the warm water in the cup is 8400 J
- the energy gained by the melted ice warming from 0°C to 14°C is about 1300 J

By equating the energy gained by the ice melting, plus the energy gained warming from 0°C to 14°C, to the energy lost by the water in the cup, show that the specific latent heat of fusion of water is about $3 \times 10^5 \, \text{J kg}^{-1}$.

> **Knowledge check 82**
>
> State the principle of conservation of energy.

> **Exam tip**
>
> Remember that the melted ice will require energy to warm it up from its melting point to the final temperature as well as the energy to melt it.

e Determine the percentage difference between this experimental value and the accepted value of $3.34 \times 10^5\,\mathrm{J\,kg^{-1}}$.

f What is probably the main reason for the experimental value being less than the accepted value?

Answer

a An expanded polystyrene cup:

■ is a good insulator

■ has a small mass

This means it will absorb very little energy, which can therefore be ignored.

b The expanded polystyrene cup is a good insulator, which reduces energy gain or loss. The technique used further minimises energy gain or loss as, to a good approximation, the energy gained by the water while it is below room temperature is balanced out by the energy lost while it is above room temperature.

c Larger pieces of ice will probably be below 0°C in the centre and so will require extra energy to bring *all* the ice to melting point. This would give a value for the latent heat that is too large.

d Energy lost by water in cup

$$= mc\Delta\theta = 0.200\,\mathrm{kg} \times 4200\,\mathrm{J\,kg^{-1}\,K^{-1}} \times (24 - 14)\,\mathrm{K} = 8400\,\mathrm{J}$$

Energy gained by the melted ice warming from 0°C to 14°C

$$= 22.6 \times 10^{-3}\,\mathrm{kg} \times 4200\,\mathrm{J\,kg^{-1}\,K^{-1}} \times (14 - 0)\,\mathrm{K} = 1329\,\mathrm{J} \approx 1300\,\mathrm{J}$$

Energy gained by ice melting + warming to 14°C = energy lost by water in cup

$$L_f \times 22.6 \times 10^{-3}\,\mathrm{kg} \quad + \quad 1329\,\mathrm{J} \quad = \quad 8400\,\mathrm{J}$$

Hence:

$$L_f = \frac{(8400 - 1329)\,\mathrm{J}}{22.6 \times 10^{-3}\,\mathrm{kg}} = 3.1 \times 10^5\,\mathrm{J\,kg^{-1}}$$

e Percentage difference $= \dfrac{(3.34 - 3.1) \times 10^5\,\mathrm{J\,kg^{-1}}}{3.34 \times 10^5\,\mathrm{J\,kg^{-1}}} \times 100\% = 7\%$

f The most probable reason for the experimental value being low is that the ice may not be dry. If some of the ice has already melted, it will not need any latent heat.

Knowledge check 83

Show that the energy needed to raise the temperature of 340 g of water from 20°C to 90°C is about 0.1 MJ.

Exam tip

Remember that, when an experimental value is being compared with a known value, the difference in values is divided by the *known value* to find the percentage difference.

Experiment B: Electrical method for latent heat of vaporisation

Planning

An elegant method uses the arrangement shown in Figure 62.

Figure 62

Knowledge check 84

Show that it will take a 2.0 kW kettle about 4 minutes to boil off 0.20 litres of water once the water has reached its boiling point. [L_v for water is 2.3 MJ kg^{-1}]

The simple procedure, without the data logger and computer, is as follows:

■ Switch on the heater of known wattage and record the balance reading at 30 s intervals.
■ Plot a graph of the balance reading m on the y-axis against the time t on the x-axis.

Safety

Ensure that you take the following precautions:

■ There is sufficient water to keep the heater fully immersed for the whole of the experiment.
■ The beaker is secure so that it does not topple.
■ Water is not spilled on the electrical equipment.
■ Electrical equipment is not touched with wet hands.

Worked example

A typical graph is shown in Figure 63.

a Explain the shape of the graph.

b Dividing the equation $\Delta E = L\Delta m$ by Δt on both sides gives us

$$\frac{\Delta E}{\Delta t} = L\frac{\Delta m}{\Delta t}$$

or, for the graph in question,

$$\frac{dE}{dt} = -L\frac{dm}{dt}$$

where dE/dt is the power of the heater (500 W) and dm/dt is the gradient of the graph.

Explain the significance of the minus sign in the equation.

c Determine the gradient of the graph and hence a value for the specific latent heat of vaporisation of water.

d Explain the most likely reason for the value obtained being larger than the accepted value of 2.34×10^6 J kg^{-1}.

e Discuss the advantages of using a computer to collect and analyse the data.

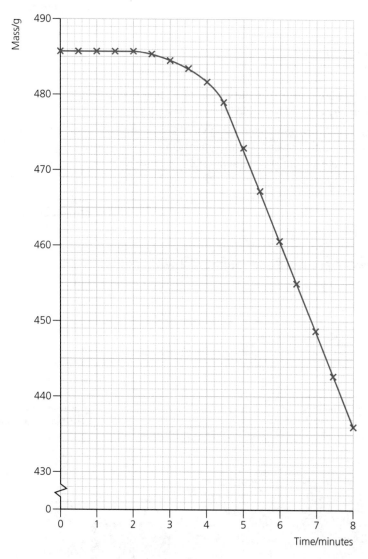

Figure 63

Exam tip

It is interesting that this 'heat' experiment does not necessitate the use of a thermometer! As school or college thermometers are notoriously inaccurate, this is a distinct advantage.

Answer

a The mass remains constant while the water is being heated to its boiling point (horizontal line), and then starts to decrease when the water begins to boil. When the water is being boiled off at a steady rate, the graph becomes linear.

b The minus sign in the equation signifies that the mass is *decreasing* with time.

c Taking a gradient between the points (4.8 min, 475 g) and (8.0 min, 436 g), converting minutes to seconds and g to kg, gives us:

$$\text{gradient} = \frac{(0.436 - 0.475)\,\text{kg}}{(480 - 288)\,\text{s}} = -2.03 \times 10^{-4}\,\text{kg s}^{-1}$$

Content Guidance

Hence:

$$L_v = \frac{500 \text{ J s}^{-1}}{2.03 \times 10^{-4} \text{ kg s}^{-1}} = 2.46 \times 10^6 \text{ J kg}^{-1}$$

d The most likely reason for this value being high is that the beaker of boiling water, despite the lagging, will be losing energy to the surroundings (at a constant rate, as its temperature above the surroundings is constant once it is boiling).

e A computer could sample at a greater rate than every 30 s and plot a graph directly. It could also be programmed to work out the gradient (and even the value of L_v).

Core practical 14

Investigating the relationship between pressure and volume of a gas at fixed temperature

Essential theory

You should be familiar with Boyle's law — the pressure of a *fixed mass* of gas is inversely proportional to its volume *provided that the temperature is kept constant*.

We can express Boyle's law mathematically by:

$$p \propto \frac{1}{V} \Rightarrow p = \text{constant} \times \frac{1}{V} \Rightarrow pV = \text{constant}$$

Planning

Two possible ways of doing the experiment are shown in Figure 64.

Figure 64

In Figure 64(a), the tube has a uniform bore, so $V \propto l$. In Figure 64(b), the volume of air can be read off directly from the calibrated syringe. In either experiment:

- as wide a range of values for p and V (or l) as possible should be recorded (but see 'Safety' below)
- the changes in pressure should be made *slowly*, with time allowed for thermal equilibrium to be reached

Safety

- With apparatus (a), it important to avoid excessive pressure and ensure connections are secure to avoid oil being sprayed at high pressure over observers.
- The apparatus should be checked by a teacher or technician before use.

Worked example

Table 16 shows a typical set of data.

Table 16

p/kPa	100	120	140	160	180	200	220	240
V/cm³	22.6	18.7	16.1	14.0	12.4	11.1	10.2	9.4
(1/V)/10³ m⁻³	44.2							

a State the independent, dependent and control variables in this experiment.

b Explain how the necessary conditions for Boyle's law are ensured.

c How could you, without drawing a graph, quickly check whether the results confirmed Boyle's law? Illustrate your answer with appropriate data.

d Complete the table by adding values of $1/V$. Note that the units are $10^3\,\mathrm{m}^{-3}$. The first value has been done for you.

e Plot a graph of p on the y-axis against $1/V$ on the x-axis and discuss whether it confirms Boyle's law.

Answer

a If you change the pressure and then measure the subsequent change in volume, the independent variable is the pressure and the dependent variable is the volume. In either case, the control variables are the mass and temperature of the air.

b To ensure the necessary conditions for Boyle's law:
  - the air in the tube or syringe must not be able to leak out, so that its mass will remain constant
  - the changes in pressure should be made *slowly*, allowing time for thermal equilibrium to be reached, so that the temperature of the air remains constant

c Try multiplying the values of pressure and volume in the table to see if they are equal.
  In kPa cm³ the values of pV are, to 3 SF: 2260, 2240, 2250, 2240, 2230, 2220, 2260.
  Within likely experimental error, these can be considered equal, thereby confirming Boyle's law.

Exam tip

You need to be confident in unit conversions, e.g. $\mathrm{cm}^3 = (10^{-2}\,\mathrm{m})^3 = 10^{-6}\,\mathrm{m}^3$.

Knowledge check 86

Sketch graphs of p against V and pV against p for the data in Table 16. You need not use graph paper, but your axes should be labelled with numerical values. Show that the unit of pV is joules.

d The remaining values of $1/V$ in $10^3\,\text{m}^{-3}$ are: 53.5, 62.1, 71.4, 80.6, 90.1, 98.0 and 106.4.

e If you plot a graph of p on the y-axis against $1/V$ on the x-axis, you should get a straight line *through the origin*. This shows that $p \propto 1/V$, which confirms Boyle's law.

Core practical 15

Investigating the absorption of gamma radiation by lead

Essential theory

You should know that, of the three types of radiation, gamma (γ) radiation is the most penetrating. While alpha (α) particles can be absorbed by a sheet of paper and beta (β) particles by a few millimetres of aluminium, γ-rays are not entirely absorbed even by several centimetres of lead. Because absorption of γ-rays is governed by probability — either the quantum of γ radiation is absorbed by an atom of the material or it is not — the probability of absorption is proportional to the thickness x. This gives an exponential absorption of the form:

$$A = A_0 e^{-\mu x}$$

where A is the activity, or count rate, and μ is a constant — a property of the absorbing material called its absorption coefficient.

Planning

The apparatus is set up as shown in Figure 65. Initially, the cobalt-60 source should not be in the laboratory.

Cobalt-60 source   Lead absorber   Geiger-Müller (G-M) tube   Counter

Figure 65

The procedure is then as follows:

■ Measure the thickness of each of the lead absorbers.
■ Determine the background count rate.
■ Put the cobalt-60 source in place.
■ Record the count rate for different thicknesses x of absorber.
■ Subtract the background count rate to get the corrected count rate A.

Safety

Danger: radiation can kill! The following safety precautions *must always be adopted* when using radioactive sources.

- Keep as far away as possible (at least 30 cm, except when handling sources—see below) from all laboratory sources of ionising radiation.
- Do not touch radioactive materials — use a handling tool.
- Keep sources in their lead storage containers, out of the laboratory, when not in use.
- During an investigation, keep the source pointed away from the body, especially the eyes.
- Limit the time of use of sources — return to secure storage as soon as possible.
- Wash your hands after working with a radioactive source.

Exam tip

You should be aware of the dangers of radiation and must learn the essential safety precautions.

Worked example

Table 17 shows a typical set of data. The count rate has been corrected for background radiation. Four lead discs of different thickness were used.

Table 17

x/mm	0.0	1.7	3.4	5.1	6.8	8.5	10.2	11.9	13.6	15.3	17.0
A/min⁻¹	405	365	340	290	270	250	217	201	189	158	150
ln(A/min⁻¹)	6.00	5.90									

a Describe how you would determine the background count rate.

b In this experiment, what is the control variable that must be kept constant?

c Deduce the thicknesses of the four discs and explain how the other values of x are obtained. Illustrate your answer with reference to $x = 13.6\,mm$.

d How would you measure the thicknesses of the four discs?

e Explain why a graph of $\ln(A/min^{-1})$ on the y-axis against x/mm on the x-axis should be a straight line.

f Complete the table to show the remaining values of $\ln(A/min^{-1})$ and then plot a graph of $\ln(A/min^{-1})$ on the y-axis against x/mm on the x-axis.

g Use your graph to determine a value for the absorption coefficient μ.

h Suggest why there is considerable scatter in the values, particularly for larger thicknesses of absorber.

Answer

a The background count should be recorded for at least 5 minutes, repeated and averaged.

b The control variable is the distance between the source and the detector, which must be kept constant.

c The thicknesses of the four discs must be the first four values of x/mm: 1.7, 3.4, 5.1 and 6.8.

The other values of x are obtained by using two or more discs in combination. For example, for 13.6 mm, a combination of the 1.7 mm + 5.1 mm + 6.8 mm discs would be used.

Exam tip

You need to be confident in obtaining a linear graph of an exponential expression by taking natural logarithms ('ln' to base 'e').

Knowledge check 88

Draw a graph of A against x for the data in Table 17. Use your graph to determine the 'half thickness', i.e. the thickness of lead needed to reduce the activity by half. What thickness would be needed to reduce the activity to a quarter of its original value?

d The thickness of each disc should be found using either a vernier calliper or a micrometer. After checking for zero error, at least three readings for each disc should be taken and averaged.

e By taking logs to base 'e' ('ln') of both sides of the equation $A = A_0 e^{-\mu x}$, we get:

$$\ln A = \ln A_0 - \mu x$$

or rearranging:

$$\ln A = -\mu x + \ln A_0$$

So a graph of $\ln (A/\text{min}^{-1})$ against x/mm should be a straight line of gradient $-\mu$ and intercept $\ln (A_0/\text{min}^{-1})$.

f The remaining values of $\ln (A/\text{min}^{-1})$ are 5.83, 5.67, 5.60, 5.52, 5.38, 5.30, 5.24, 5.06 and 5.01. Your graph should look like Figure 66.

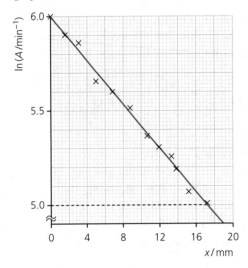

Figure 66

> **Exam tip**
>
> Note that, as the origin has no particular significance, the y-scale of the graph starts at 5.0 to give a large scale. This is often the case when plotting log graphs. Here, a graph that included the origin would be penalised.

g From the points (17.2, 5.00) and (0.0, 6.00):

$$\text{gradient} = -\frac{5.00 - 6.00}{(17.2 - 0.0)\,\text{mm}} = -0.058\,\text{mm}^{-1} = -\mu$$

If you draw lines of maximum and minimum slope, you will find that the uncertainty in the gradient is about 0.002, giving a value for μ of $0.058 \pm 0.002\,\text{mm}^{-1}$.

h The scatter is most likely to be due to the random nature of radioactivity. With larger thicknesses, the count rate is smaller and so the random effect is more apparent.

Core practical 16

Determining the value of an unknown mass using the resonant frequencies of the oscillation of known masses

Essential theory

Although the specification states 'using the resonant frequencies', this is, perhaps, not quite the right wording. What is required is timing the vertical oscillations of a mass on a spring and hence finding the *natural frequency* of oscillations. Although, for undamped oscillations, the resonant frequency has the same value as the natural frequency, in the simple situation of a mass oscillating on a spring, there is no driving force, and so this is not actually 'resonance'.

You need to recall that a mass m oscillating on a light vertical spring that obeys Hooke's law and has a spring constant k will execute simple harmonic motion. The time period of the oscillations is given by:

$$T = 2\pi\sqrt{\frac{m}{k}}$$

Knowledge check 89

Show that rearranging the equation $T = 2\pi\sqrt{\frac{m}{k}}$ gives $T^2 = \frac{4\pi^2}{k} \times m$.

Planning

You should note the following:

- The experiment works well with a standard 20 mm, so-called 'expendable', laboratory spring — 'so-called' because your teacher may not be best pleased if you 'expend it'! You should therefore be careful not to stretch it beyond its elastic limit.
- You also need a range of 100 g masses and a 50 g mass, so that the spring can be loaded up to about 400 g in intervals of 50 g.
- A convenient 'unknown' mass could be a block of wood having a volume of about 500 cm^3.
- A stopwatch is needed for timing.

Safety

See question **b** in the worked example below.

Worked example

Table 18 shows a typical set of data.

Table 18

m/kg	0.100	0.150	0.200	0.250	0.300	0.350	0.400	Block
T/s	0.410	0.502	0.574	0.648	0.710	0.767	0.819	0.663
T²/s²	0.168	0.252						

a Draw a diagram of how you would set up the apparatus.

b State any safety precautions that you would observe in conducting the experiment.

c Suggest why a block of wood having a volume of about 500 cm^3 might be a convenient 'unknown' mass.

Content Guidance

d Explain the experimental techniques you would use to obtain data such as those in Table 18, adding to your diagram if necessary.

e Discuss why using a graph of T^2 on the y-axis against m on the x-axis is a better way of determining the mass of the wooden block than by plotting a graph of T against m.

f Complete Table 18 and plot a graph of T^2 against m. Use your graph to determine the mass M of the block of wood.

g The block was found to measure $(89 \pm 1\,\text{mm}) \times (75 \pm 1\,\text{mm}) \times (75 \pm 1\,\text{mm})$. Use these data and your mass for the block to determine the density of the wood. Estimate the uncertainty in your value for the density, explaining how you made your estimate.

> **Exam tip**
> You need to be able to make sensible estimates of quantities, such as the density of wood being of the order of, perhaps, 500–600 kg m⁻³ (since wood floats).

Answer

a Figure 67 shows how the apparatus should be arranged.

Figure 67

b A shock pad should be placed under the masses, and the stand should be secured so that it does not topple when the masses are added to the spring.

c Assuming the density of wood is about $500\,\text{kg m}^{-3}$, a volume of $500\,\text{cm}^3$ $(5 \times 10^{-4}\,\text{m}^3)$ would have a mass of about $0.25\,\text{kg}$, which would be in the middle region of the graph.

d The following techniques should be used:
- Make sure that the spring is gripped firmly so that there is no slipping or twisting.
- Time, say, 20 *small* oscillations and repeat. Average and divide by 20 to find T. The oscillations should be small so that the spring is not stretched beyond the Hooke's law region.
- Use a marker at the *centre* of the oscillations to aid counting, as shown in Figure 67.

> **Exam tip**
> A marker ('fiducial' mark) at the *centre* of the oscillations should always be used when timing oscillations — and do not forget to time a *large number* of oscillations and then *divide by the number of oscillations* to find the period. These are points that candidates often omit and consequently lose marks.

Figure 67 labels: Spring gripped firmly between two pieces of wood; Small vertical oscillations; Pin at centre of oscillations; Eye level with pin; Blu-Tack e A graph of T^2 against m will give a straight line, while a graph of T against m will be a curve. As it is much easier to draw a line of best fit for points that lie on a straight line, a graph of T^2 against m is better.

f The remaining values of T^2/s^2 are 0.329, 0.420, 0.504, 0.588, 0.671 and 0.440. Your graph should be a straight line through the origin which, by reading off the mass corresponding to $T^2 = 0.440\,s^2$, gives the mass M of the block as 0.260 kg.

g The volume of the block is $(89 \times 75 \times 75) \times 10^{-9}\,m^3 = 5.0 \times 10^{-4}\,m^3$. So:

$$\text{density} = \frac{\text{mass}}{\text{volume}} = \frac{0.260\,\text{kg}}{5.0 \times 10^{-4}\,\text{m}^3} = 520\,\text{kg}\,\text{m}^{-3}$$

The percentage uncertainty in the volume is $(1/89 + 1/75 + 1/75) \times 100\% = 3.8\%$. The mass can be read off the graph to about ± 0.002 kg, giving a percentage uncertainty of $(0.002/0.260) \times 100\% = 0.8\%$ (or we could say that the timing uncertainty for the block is human error of 0.1 s in a time of about 13 s for 20 oscillations of the block — this also gives a percentage uncertainty of about 0.8%). The percentage uncertainty in the density is therefore $3.8\% + 0.8\% = 4.6\%$, giving an uncertainty of:

$4.6\% \times 520\,\text{kg}\,\text{m}^{-3} = \pm 24\,\text{kg}\,\text{m}^{-3}$

The density of the wood is probably best expressed as $520 \pm 20\,\text{kg}\,\text{m}^{-3}$ (i.e. somewhere between $500\,\text{kg}\,\text{m}^{-3}$ and $540\,\text{kg}\,\text{m}^{-3}$), as the measurements of the block were only to 2 SF.

Knowledge check 90

Use the gradient of your graph of T^2 against m for the data in Table 18 to determine a value for the spring constant k of the spring.

Questions & Answers

The Edexcel examinations

The Edexcel A-level Physics examination consists of three papers. All papers will examine 'Working as a Physicist'. Briefly, this means students:

- working scientifically, developing competence in manipulating quantities and their units, including making estimates
- experiencing a wide variety of practical work, developing practical and investigative skills by planning, carrying out and evaluating experiments and becoming knowledgeable of the ways in which scientific ideas are used
- developing the ability to communicate their knowledge and understanding of physics
- acquiring these skills through examples and applications from the entire course

In particular, paper 3 covers the general and practical principles of physics. It is of 2 hours and 30 minutes duration and has 120 marks. This paper may draw on any of the topics in the specification and includes:

- synoptic questions that may draw on two or more different topics
- questions that assess conceptual and theoretical understanding of experimental methods (indirect practical skills) that will draw on students' experiences of the core practicals

This student guide covers only practical work, mainly through consideration of the 16 core practicals. Therefore, the questions in this section only reflect the type of question that will be set in paper 3 to test your understanding of experimental methods and are *not* representative of the paper as a whole. Questions on practical work may also be asked in papers 1 and 2.

In the Edexcel AS Physics examination, there are just two papers, and questions based on practical work may be set in both papers.

A formulae sheet is provided with each test. Copies may be downloaded from the Edexcel website, or can be found at the end of past papers.

In addition to the written papers, for A-level there is a Science Practical Endorsement, for which you will be assessed separately. The Endorsement will not contribute to the overall grade for your A-level qualification, but the result will be recorded on your certificate. In order to gain the Endorsement, you need to provide evidence that you have completed the necessary practical activities as indicated in the Edexcel specification. You must complete a minimum of 12 identified practical activities that demonstrate your competence to:

- follow written procedures
- apply investigative approaches and methods when using instruments and equipment
- safely use a range of practical equipment and materials
- make and record observations
- research, reference and report

You should refer to the Edexcel specification, where there is a table in Appendix 5d showing how each core practical maps to the required practical techniques.

Examiners use certain **command terms** that require you to respond in a particular way, for example 'state', 'explain' or 'discuss'. You must be able to distinguish between these terms and understand exactly what each requires you to do. A full list can be found in Appendix 7 in the Edexcel A-level specification or Appendix 4 of the AS specification. You are strongly recommended to download this (www.edexcel.com) and print off a copy for reference.

You should pay particular attention to drawing diagrams, plotting graphs and making calculations. Many candidates lose marks by failing to label diagrams properly, not giving essential data on graphs and, in calculations, by not showing all the working or by omitting units.

The answers that follow should not be treated as model answers — they represent the bare minimum necessary to gain the marks. Some questions have tips on how to answer the question, preceded by **e**. Comments denoted by **e** may either provide you with useful tips or indicate where candidates often lose marks. Ticks (✓) are included in the answers to show where a mark has been awarded.

∎ Practice questions

Multiple choice

Question 1

A student records the diameter of a wire five times using a micrometer reading to 0.01 mm. The readings are 1.17 mm, 1.19 mm, 1.17 mm, 1.16 mm and 1.18 mm. The student averages these readings, but does not take into account the zero error on the micrometer. The average measurement of the diameter is:

A precise and accurate

B precise but not accurate

C accurate but not precise

D not accurate and not precise (1 mark)

Question 2

A student is measuring the force on a current-carrying conductor using the apparatus shown in Figure 1. She measures the length of the copper wire in the magnetic field using a 30 cm rule, which has markings every millimetre.

The uncertainty in this measurement of length is:

A ±1 cm   **B** ±0.5 cm   **C** ±0.2 cm   **D** ±0.05 cm

(1 mark)

Figure 1

Question 3

In an experiment to determine the specific latent heat of vaporisation of water, steam is condensed in water at room temperature. A student records the initial and final temperatures of the water as $\theta_i = (19.0 \pm 0.5)°C$ and $\theta_f = (29.0 \pm 0.5)°C$. The percentage uncertainty in the temperature difference $\Delta\theta$ is:

A 3%   **B** 5%   **C** 10%   **D** 20%   (1 mark)

Question 4

An object falls from rest with an acceleration g. The variation with time t of the displacement s of the object is given by $s = \frac{1}{2}gt^2$. The uncertainty in the value of t is ±3% and the uncertainty in the value of s is ±2%. The best estimate for the uncertainty of the value of g is:

A 5%   **B** 6%   **C** 8%   **D** 11%   (1 mark)

Question 5

A student measures the wavelength of a laser using a diffraction grating having 300 lines per millimetre. The diffraction pattern is observed on a white screen placed at a distance of 0.5 m from the grating. The first order maxima are a distance x apart. The grating is replaced by one having 600 lines per millimetre, and the screen is moved back so that it is 1.0 m from the grating. The distance apart of the first order maxima is now approximately:

A $x/2$ **B** x **C** $2x$ **D** $4x$ (1 mark)

Question 6

(A-level only) A student is investigating the absorption of gamma rays by lead. With no absorber, the count rate, when corrected for background, is $400\,min^{-1}$. When a lead disc is placed between the source and detector, the count rate falls to $300\,min^{-1}$. If a second, identical, disc is added the count rate will drop to:

A $150\,min^{-1}$ **B** $200\,min^{-1}$ **C** $225\,min^{-1}$ **D** $250\,min^{-1}$ (1 mark)

Question 7

A student is investigating how the frequency of a stretched wire depends on the mass per unit length μ of the wire, using wires of different diameter. The tension and length of each wire are kept constant. The student plots a graph of $\ln f$ on the y-axis against $\ln \mu$ on the x-axis. The gradient will be:

A $-\dfrac{1}{\sqrt{2}}$ **B** $-\dfrac{1}{2}$ **C** $\dfrac{1}{2}$ **D** $\dfrac{1}{\sqrt{2}}$ (1 mark)

Answers to Questions 1–7

1 B

ⓔ The repeated readings are close to each other, which means that they are *precise*. The student neglected to take into account the zero error and therefore each reading may be larger or smaller than the true reading. The readings are therefore *not accurate*.

2 C

ⓔ As the measurement of length will be the difference of two readings, each to an uncertainty of 0.1 cm, the uncertainty in the length will be 2 × 0.1 cm = 0.2 cm. (The uncertainty of each reading could be taken as 0.05 cm, giving an overall uncertainty of 0.1 cm, but that is not one of the possible answers.)

3 C

ⓔ When a value is found by adding or subtracting two quantities, the maximum and minimum possible values have to be calculated to find the uncertainty in the value, and then the percentage uncertainty can be calculated.

4 C

ⓔ Remember to add the percentage uncertainties if quantities are divided; and if a quantity is raised to a power, multiply the percentage uncertainty by the power. Then: % uc in g = % uc in s + 2 × % uc in t.

5 D

ⓔ Use $n\lambda = d\sin\theta$. Practise using ratios, as they are common in multiple-choice questions. Remember, for small angles $\sin\theta \approx \tan\theta \approx \theta$ (in radians).

6 C

ⓔ With one disc, the count rate reduces to 300 min^{-1} = 0.75 × 400 min^{-1}. So, for two discs, the count rate will decrease to 0.75 × 300 min^{-1} = 225 min^{-1}. This is a quick way of checking whether a graph is exponential.

7 B

ⓔ From $f = \dfrac{1}{2l}\sqrt{\dfrac{T}{\mu}} \Rightarrow f = \dfrac{\sqrt{T}}{2l} \times \dfrac{1}{\sqrt{\mu}} = \dfrac{\sqrt{T}}{2l} \times \mu^{-1/2} \Rightarrow \ln f = \ln\left(\dfrac{\sqrt{T}}{2l}\right) - \dfrac{1}{2}\ln\mu$

You need to practise taking logs of equations such as this. Although you could also take logs to base 10 ('log' on your calculator) when trying to find a power, if you *always* take natural logs ('ln'), you will not get the wrong logs when dealing with exponential functions (which *must* be 'ln').

Structured questions

Question 8

A rectangular glass block is measured with a ruler calibrated in 1 mm divisions:

      length = 11.5 cm   width = 6.2 cm   depth = 3.1 cm

(a) Calculate the volume of the block.   (1 mark)

(b) Estimate the percentage uncertainty and hence the absolute uncertainty in your value.   (3 marks)

(c) State how the volume of the block should be recorded.   (1 mark)

          Total: 5 marks

Student answer

(a) volume = 11.5 cm × 6.2 cm × 3.1 cm = 221.03 cm^3 = 221 cm^3 ✓

(b) % uc in volume = % uc in length + % uc in width + % uc in depth

   = (0.1 cm/11.5 cm) × 100% + (0.1 cm/6.2 cm) × 100% + (0.1 cm/3.1 cm) × 100% ✓

   = (0.870 + 1.61 + 3.23)% = 5.7% ✓

   Absolute uncertainty in volume = 5.7% of 221 cm^3 = 13 cm^3 ✓

🄴 Remember that the percentage uncertainties are added when quantities are multiplied.

(c) Volume of block = (221 ± 13) cm^3 ✓

🄴 Although the width and depth are only measured to 2 SF, an answer of 221 ± 13 cm^3 is arguably better than giving the answer as 220 ± 10 cm^3 to 2 SF. Full credit would be given for either answer.

Question 9

A student is measuring the e.m.f. generated by a magnet falling through a coil of wire using a data logger and voltage sensor.

(a) Explain why using a data logger and voltage sensor is a good strategy.   (3 marks)

(b) After collecting her data, she realises that the sensor has a zero error. Explain what is meant by a zero error, and describe what she should do with her data to take the error into account.   (2 marks)

          Total: 5 marks

Student answer

(a) The time for the magnet to fall through the coil is very short. ✓

   As a data logger can sample at a high rate, ✓ it can take a large number of readings in this short time. ✓

e As the question says 'explain', it is necessary to state that the time involved is small.

> **(b)** Zero error means that the voltage sensor gives a non-zero reading when there is no measurement being taken. ✓
>
> She should subtract the value of the non-zero reading from her e.m.f. data. ✓

Question 10

(A-level only) A student used the arrangement shown in Figure 2 to measure the specific heat capacity of water.

Figure 2

(a) State the measurements the student would need to make in order to measure the specific heat capacity of water. (2 marks)

(b) Explain how the student could try to make the measurements as accurate as possible. (4 marks)

(c) State *one* safety precaution you would take if you were to carry out this experiment. (1 mark)

Total: 7 marks

> **Student answer**
>
> **(a)** You would need to calculate the energy supplied electrically by measuring the current, p.d. and time of heating ✓ (or electrical energy supplied $W = VIt$ ✓).
>
> You would also need to measure the mass and the initial and final temperatures of the water. ✓
>
> **(b)** Points might include the following:
> - Stir the liquid regularly to ensure that it is all at the same temperature.
> - Record the final steady temperature after switching off to allow equilibrium to be attained.
> - Determine the amount of energy absorbed by the container and deduct this from the energy supplied by the heater. This would give you the actual energy supplied to the liquid.
> - Insulate the container to prevent energy being lost to the surroundings.
> - Leave the thermometer in the liquid to ensure it is at thermal equilibrium with the water and ensure it is not touching the walls of the container.
>
> Any two points ✓✓ explained ✓✓

ⓔ As the question asks you to 'explain', you must both state the technique *and* explain why it improves the accuracy. Note that using an expanded polystyrene beaker both provides insulation *and* avoids having to determine any energy taken by the container.

(c) Safety precautions could include:
- Make sure the wiring does not short circuit.
- Stand up at all times and clamp the beaker to avoid scalding in case the beaker of hot water is knocked over.
- Keep the immersion heater fully immersed.

Any one ✓

Question 11

Figure 3 shows an oscilloscope trace of a sound wave. The signal has a frequency of 2.5 kHz and a peak voltage of 500 mV.

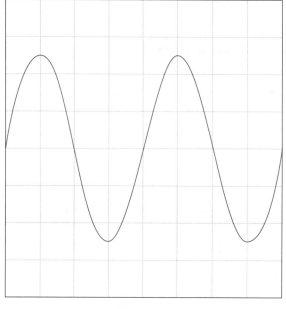

Figure 3

(a) What are the time base and voltage settings on the oscilloscope? (3 marks)

ⓔ This question is based on core practical 6. You are expected to be familiar with the use of an oscilloscope.

(b) The speed of sound is 330 m s⁻¹. What is the wavelength of this sound wave? (1 mark)

(c) (A-level only) Determine the root-mean-square voltage displayed on the oscilloscope. (2 marks)

Total: 6 marks

Student answer

(a) $T = 1/f = 1/(2500\,s^{-1}) = 4.00 \times 10^{-4}\,s$ ✓ $= 400\,\mu s$

$T =$ time for one wavelength = four divisions $\Rightarrow$ time base = $100\,\mu s$/div ✓

$V_{peak} = 500\,mV = 2.5$ divisions $\Rightarrow$ voltage setting = $500\,mV/2.5\,div = 200\,mV$/div ✓

ⓔ To get full marks, you would be expected to express the time base in realistic units — either (preferably) μs/div or ms/div and not 10^{-4} s/div.

(b) $c = f\lambda \Rightarrow \lambda = c/f = 330\,m\,s^{-1}/2500\,s^{-1} = 0.132\,m$ (or 13 cm) ✓

(c) $V_{rms} = \dfrac{V_{peak}}{\sqrt{2}}$ ✓ $\Rightarrow V_{rms} = \dfrac{500\,mV}{\sqrt{2}} = 350\,mV$ ✓

Question 12

(A-level only) A student is investigating the properties of a spring. He finds that a mass of 300 g suspended from the spring causes it to extend by 120 mm. When he pulls the mass down a small distance and releases it, he times 20 oscillations in 14.12 s using a digital stopwatch.

(a) Use these data to determine *two* values for the spring constant k of the spring. (4 marks)

(b) Show that the unit $N\,m^{-1}$ is equivalent to $kg\,s^{-2}$. (1 mark)

(c) Calculate the percentage difference between your two values and discuss which value is likely to have the greater uncertainty. (4 marks)

(d) Explain why it is important to pull the mass down only a small distance. (2 marks)

Total: 11 marks

Student answer

(a) $F = kx \Rightarrow k = F/x = (0.300\,kg \times 9.81\,N\,kg^{-1})/0.120\,m$ ✓ $= 24.5\,N\,m^{-1}$ ✓

$T = 2\pi\sqrt{\dfrac{m}{k}} \Rightarrow k = \dfrac{4\pi^2 m}{T^2} = \dfrac{4\pi^2 \times 0.300\,kg}{(0.706\,s)^2}$ ✓ $= 23.8\,kg\,s^{-2}$ ✓

ⓔ Marks are often lost by either not squaring π or by forgetting to divide the time by the number of oscillations to find T.

(b) $N\,m^{-1} = (kg\,m\,s^{-2}) \times m^{-1} = kg\,s^{-2}$ ✓

(c) percentage difference $= \dfrac{24.5 - 23.8}{24.2} \times 100\% = 3\%$ ✓

The uncertainty in measuring the extension is likely to be 2 mm at the most, giving a percentage uncertainty of (2 mm/120 mm) × 100% = 1.7%. ✓

Although the time is recorded to 0.01 s, in reality, the uncertainty is governed by human reaction time, say 0.1 s, giving a percentage uncertainty in T of (0.1 s/14.12 s) × 100% = 0.7%. ✓

The percentage uncertainty in T^2 will therefore be 2 × 0.7% = 1.4%. ✓

It would therefore appear that the two methods are comparable in terms of uncertainty.

ⓔ It is important that you can make sensible estimations of uncertainties. A common mistake here would be to quote the uncertainty in the timing as 0.01 s, when reaction time is likely to be 0.1 s. Remember, the percentage uncertainty is *doubled* if the quantity is *squared*, and the *average* value should be used in the denominator when determining the percentage difference between two *experimental* values.

> **(d)** For simple harmonic motion, the acceleration and, therefore, the force must be proportional to the displacement. ✓ The spring should be given a *small* displacement to ensure that it oscillates within the Hooke's law region, so that force ∝ displacement. ✓

Question 13

Figure 4 shows an arrangement that a student sets up for investigating the acceleration of a glider on a tilted air track. For clarity, the lamps illuminating the light gates have been omitted.

Figure 4

(a) The glider is released from rest and the interrupter card, of length 150 mm, is timed as it passes through the two light gates. Times of t_1 = 1.49 s and t_2 = 0.24 s are recorded.

    (i) Calculate the average velocities v_1 and v_2 of the glider as it passes through each light gate.

    (ii) The light gates are 200 mm apart. Calculate the acceleration of the glider.

ⓔ Using $v^2 = u^2 + 2as$ instead of $v = u + at$ is an alternative way of determining an acceleration when investigating motion.

    (iii) Show that the theoretical acceleration of the glider is $g \sin \theta$, where θ is the angle that the track makes with the horizontal.

    (iv) The student measures the angle by means of a protractor. She records a value of 6.0° ± 0.5°. Determine a value for g from the data recorded by the student.

    (v) Estimate the uncertainty in this value for g. What conclusion can you come to regarding the validity of this experiment?

    (vi) Suggest why the value of acceleration obtained is likely to be too small.   (10 marks)

(b) The student's teacher suggests that a much better way to determine θ would be by measuring the distances h and L as shown in Figure 5. The student records $h = 98 \pm 1$ mm and $L = 1000 \pm 1$ mm. What value does this give for θ? Comment on this value.

(4 marks)

(c) Explain how you could extend this experiment to get a more reliable value for g. Include a sketch of the graph you would plot and how you would find g from it. Explain how your graph would reduce both systematic and random errors.

(6 marks)

Total: 20 marks

Student answer

(a) (i) $v_1 = 0.150\,\text{m}/1.49\,\text{s} = 0.101\,\text{m s}^{-1}$ and $v_2 = 0.150\,\text{m}/0.24\,\text{s} = 0.625\,\text{m s}^{-1}$ ✓

(ii) $v^2 = u^2 + 2as \Rightarrow a = \dfrac{v_2^2 - v_1^2}{2s} \Rightarrow a = \dfrac{(0.625\,\text{m s}^{-1})^2 - (0.101\,\text{m s}^{-1})^2}{2 \times 0.200\,\text{m}}$ ✓

$a = 0.95(1)\,\text{m s}^{-2}$ ✓

(iii) Refer to the free-body diagram for the glider below.

Force on glider F = component of mg down the plane = $mg \sin \theta$. ✓

$F = ma \Rightarrow a = F/m = (mg \sin \theta)/m = g \sin \theta$ ✓

(iv) $a = g \sin \theta \Rightarrow g = a/\sin \theta = 0.951\,\text{m s}^{-2}/\sin 6.0° = 9.1\,\text{m s}^{-2}$ ✓

(v) $0.951\,\text{m s}^{-2}/\sin 5.5° = 9.9\,\text{m s}^{-2}$ and $0.951\,\text{m s}^{-2}/\sin 6.5° = 8.4\,\text{m s}^{-2}$ ✓

Uncertainty in $g = \pm0.8\,\text{m s}^{-2}$ ✓

The accepted value of $g = 9.8\,\text{m s}^{-2}$ just falls within this (large) uncertainty, so the value of g obtained suggests the experiment could be valid. ✓

(vi) The value obtained is likely to be too small, as there is almost certain to be some frictional force slowing down the glider. ✓

(b) $\tan \theta = h/L = 98\,\text{mm}/1000\,\text{mm} = 0.098 \Rightarrow \theta = \tan^{-1}(0.098) = 5.60°$ ✓

Largest value could be $\tan^{-1}(0.099) = 5.65°$ and smallest value $\tan^{-1}(0.097) = 5.54°$ ✓

So we can say $\theta = 5.60° \pm 0.05°$, ✓ which is a far (10 times) greater precision than using a protractor. ✓

e Choosing the right instrument, or technique, for making a measurement, and understanding its limitations, are important skills that you should develop during the course of your practical work.

(c) The angle of the track should be varied to give different values of θ (or $\tan \theta$). ✓

A graph of a against $\tan \theta$ should be plotted, ✓ which should be a straight line through the origin (sketch should show this). ✓

The gradient of this graph is g. ✓

The graph will reduce random errors by averaging several values. ✓

If there is a systematic error due to frictional forces, this will be indicated by a small intercept on the $\tan \theta$ axis, but the gradient will still be g. ✓

---

e It is important that you understand how a graph can reduce random errors and show systematic errors.

Knowledge check answers

1 **a** $kg\,m^2\,s^{-2}$, **b** As, **c** $kg\,m^2\,s^{-3}\,A^{-2}$, **d** $kg\,m^2\,s^{-3}$

2 Diameter = $1.27 \times 10^{-2}\,m$
Length = $0.800\,m$
Area = $1.27 \times 10^{-4}\,m^2$
Volume = $1.02 \times 10^{-4}\,m^3$

3 Wavelength = $6.43 \times 10^{-7}\,m$
Slit distance = $3.3 \times 10^{-6}\,m$

4 $c_w = \dfrac{c}{n_w}$ and $c_0 = \dfrac{c}{n_0}$ so $c_w : c_0 = n_0 : n_w$
$c_w : c_0 = 1.47 : 1.33 = 1.11$

5 **a** 48, **b** 300, **c** 8.0×10^3

6 $n = \sin 35.5° / \sin 26.0° = 1.32$

7

Angle	$\sin\theta$	$\cos\theta$	$\tan\theta$
$\theta = 0°$	0	1	0
$\theta = 45°$	0.707	0.707	1
$\theta = \pi\,rad$	0	−1	0
$\theta = \pi/2\,rad$	1	0	∞

8 **a** 1.08×10^{21}, **b** $3.0 \times 10^8\,ms^{-1}$, **c** 1.8×10^2 or 180

9 **a** 10^{18}, **b** 10^3, **c** 10^2, **d** 10^9

10 **a** Assume popcorn is cube of side $10^{-2}\,m$ so volume $10^{-6}\,m^3$
Assume room is cube of side $10\,m$ so volume $10^3\,m^3$
Number of pieces of popcorn = $10^3\,m^3 / 10^{-6}\,m^3 = 10^9$
b Assume you are made of water. Number of electrons per molecule is 10^1
Mass of one molecule of water is $10^{-27}\,kg$
Assume your mass is $10^2\,kg$
Number of molecules of water you contain = $10^2 / 10^{-27} = 10^{29}$
$\Rightarrow$ Number of electrons = $10^{29} \times 10^1 = 10^{30}$
c $E_p = mg\Delta h$
Assume Mount Everest is $10^5\,m$ high:
$E_p = 10^2\,kg \times 10^1\,ms^{-2} \times 10^5\,m = 10^8\,J$
Assume energy from chocolate bar = $10^5\,J$
Number of bars = $10^8 / 10^5 = 10^3$

11 **a** $W = VQ$, **b** $x \propto F$, **c** $\Phi = BA$

12
a $V = \dfrac{m}{\rho}$ **b** $m = \dfrac{p}{v}$
c $R_2 = R_1 \left(\dfrac{V_{out}}{V_{in} - V_{out}} \right)$ **d** $a = \dfrac{v^2 - u^2}{2s}$
e $\theta = \sin^{-1}\left(\dfrac{n\lambda}{d} \right)$ **f** $\omega = \sqrt{\dfrac{F}{mr}}$

13 **a** Moment = $225\,N \times 0.07\,m = 15.8\,Nm$
b $L = x/\varepsilon = 0.10\,m / 0.1 = 1\,m$
c $x = 0.040\,m \times \cos[(\pi/3)\,rad\,s^{-1} \times 11\,s]$
$x = 0.020\,m$ (remember to put your calculator in radian mode)

14 From $s = ut + \frac{1}{2}at^2 \Rightarrow t^2 = 2h/g$ (as $u = 0$)
$\Rightarrow t = 0.55\,s$

15 Acidity (pH scale)
Earthquake intensity (Richter scale)
Hardness (Mohs scale)
Star brightness (magnitude scale)
Wind intensity (Beaufort scale)

16 **a** plot λ (y-axis) against $\dfrac{1}{f}$ (x-axis); gradient = c
b plot R (y-axis) against l (x-axis); gradient = $\dfrac{\rho}{A}$
c plot T^2 (y-axis) against m (x-axis); gradient = $\dfrac{4\pi^2}{k}$

17 Instantaneous acceleration =
$(20\,ms^{-1} - 5.5\,ms^{-1})/(6.1\,s - 0.0\,s) = 2.4\,ms^{-2}$
Average acceleration = $17.5\,ms^{-1}/6.0\,s = 2.9\,ms^{-2}$

18 **a** Energy stored in a spring: $0.36\,J$ to $0.38\,J$
b Charge: $3300\,\mu C$ to $3400\,\mu C$

19 **a** graph a, **b** graph c, **c** graph d

20 Volume = $\pi r^2 h$
= $\pi \times (6.0 \times 10^{-3}\,m)^2 \times 1.50\,m = 1.7 \times 10^{-4}\,m^3$

21 Mass of one atom = molar mass/Avogadro constant
= $(12.0 \times 10^{-3}\,kg\,mol^{-1})/(6.02 \times 10^{23}\,mol^{-1})$
= $1.99 \times 10^{-26}\,kg$
Volume of one atom = mass/density
= $5.69 \times 10^{-30}\,m^3$
Assuming atom is spherical:
$V = \frac{4}{3}\pi r^3$
$r = \sqrt[3]{\dfrac{3V}{4\pi}} = 1.1 \times 10^{-10}\,m$

22 **a** $0.262\,rad$, **b** $0.873\,rad$, **c** $2.53\,rad$

23 **a** $22.5°$, **b** $85.9°$, **c** $8.59°$

24

Angle	$\sin\theta$	$\cos\theta$	$\tan\theta$
$5.0\,rad$	−0.96	0.28	−3.4
$0.50\,rad$	0.48	0.88	0.55
$0.050\,rad$	0.050	1.0	0.050

25 There are many examples of random errors where measurements are affected in an unpredictable fashion: e.g. changes in room temperature, contact resistance, trying to read off the current from an ammeter when the current is changing quickly, human reaction time when starting and stopping a stopwatch.

26 $0.86 - 0.02 = 0.84\,mm$

27 Accurate — how close to the true value a measurement is.
Precision — how close to each other repeated measurements are.

28 The range of data is large ($62\,mT - 46\,mT = 16\,mT$), giving a 16% variation either side of the mean value. The data are therefore not precise.

29 The smallest change in value that can be measured using the instrument.

30 Largest = 100 + 5% = 105 Ω
Smallest = 100 − 5% = 95 Ω

31 a Uncertainty in measurement = 0.1 mm.
Diameter = 21.5 mm ± 0.1 mm

 b Percentage uncertainty =
(0.1 mm/21.5 mm) × 100% = 0.5%

32 Inconsistent decimal places in the first column. Incorrect column headings (should be p.d./V and Current/A). Units should not appear in the body of the table. Current only recorded to 1 SF.

33 Mean a = 0.453 m s^{-2} = 0.45 m s^{-2} to 2 SF

34 a Correct reading is 8.3 cm (or 83 mm)

 b With a mirror behind the pointer, move your head until the image is hidden by the pointer. Then you are reading the scale vertically, which will give you the correct value.

35 Vernier reads 12.27 cm (or 122.7 mm)

36 Volume of sphere $V = \frac{4}{3}\pi r^3$

$$V = \frac{4}{3} \times \pi \times \left(5.0 \times 10^{-3}\,\text{m}\right)^3$$

= 5.24 × 10^{-7} m^3
Density = mass/volume
= (1.3 × 10^{-3} kg)/(5.24 × 10^{-7} m^3) = 2483 kg m^{-3}
= 2.5 × 10^3 kg m^{-3} or 2.5 g cm^{-3} to 2 SF (in line with the data)

37 a 0.5 mm (the bottom half of the main scale subdivides the 1 mm divisions of the top half of the scale into 0.5 mm)

 b 7.25 mm, assuming the micrometer reads zero when closed/there is no zero error

38 There is a positive zero error, so this must be subtracted from each reading. The actual diameter is therefore: 2.12 mm − 0.03 mm = 2.09 mm.

39 Random errors must be reduced. You should be able to analyse each source of random error in the experiment and suggest ways of reducing it. This will include taking repeats, but will also involve choosing apparatus with greater resolution or improving experimental technique.

40 Use vernier callipers (or a micrometer). Check for zero error. Then measure the diameter several times, at different angles, and average.
Radius = ½ × diameter. Use the equation:
volume of a sphere $V = \frac{4}{3} \times \pi r^3$

41 a Human reaction time is at least 0.1 s, which is far more significant than the resolution of the stopclock (typically 0.01 s) when measuring short times.

 b Uncertainty = ±0.1 s (or possibly ±0.2 s)

42 (0.1 s/18.9 s) × 100% = 0.53%

43 Use the set-square with one side on the bench to check that the vertical height of the beam above the bench is the same in different places along the length of the beam.

44 a $\sin\theta$ = 87 mm/1000 mm = 0.087 ⇒ θ = 5.0°

 b Uncertainty in lengths = (say) ±2 mm
% uc in height = (2 mm/87 mm) × 100% = 2.3%
% uc in length = (2 mm/1000 mm) × 100% = 0.2%
% uc in tan θ = 2.3% + 0.2% = 2.5%
As angle is small (< 5°) we can say tan $\theta \approx \theta$ and so:
% uc in angle = 2.5%

45 $\frac{1}{2}mv^2 = mg\Delta h \Rightarrow v = \sqrt{2g\Delta h} \Rightarrow$
$$v = \sqrt{2 \times 9.81 \times 0.10} = 1.4\ \text{m s}^{-1}$$

46 (1.44 + 1.42 + 1.46 + 1.43)/4 = 1.44
1.56 is not included in the mean calculation, as it is clearly anomalous.

47 60 s × 0.5 s^{-1} = 30 measurements

48 Using $s = ut + \frac{1}{2}at^2$ where u = 0 and a = 9.81 m s^{-2}:
$$t = \sqrt{\frac{2s}{a}} = \sqrt{\frac{2 \times 0.05\,\text{m}}{9.81\,\text{m s}^{-2}}} = 0.101\,\text{s} \approx 100\,\text{ms}$$

49 120 m = k × 2000 kg
$x = k$ × 1400 kg
x = (1400 kg/2000 kg) × 120 m = 84 m

50 a Plot v^2 on y-axis against h on x-axis. The graph should be a straight line through the origin with gradient $2g$.

 b Plot T^2 on y-axis against l on x-axis. The graph should be a straight line through the origin with gradient $4\pi^2/g$.

51 $R = V/I$ = (4.97 × 10^{-3} V)/(22.6 × 10^{-6} A) = 220 Ω

52 a For a metallic conductor, the potential difference is directly proportional to the current, provided the temperature remains constant.

 b Your sketch should have axes labelled V/V and I/A.

 c If the component is ohmic, the graph will be a straight line through the origin.

 d If V is plotted against I, the resistance will be equal to the gradient.

53 ln (8.2) − ln (8.0) = 2.10 − 2.08 = 0.02
ln (8.4) − ln (8.2) = 2.13 − 2.10 = 0.03
The point ln (8.2) = 2.10 should be plotted with an error bar between 2.08 and 2.13

54 Graph should be a straight line of negative slope. The gradient is equal to $-\lambda$ (where λ is the decay constant).
Then $t_{1/2} = \dfrac{\ln 2}{\lambda} = \dfrac{0.693}{\lambda}$.

55 a percentage difference $= \dfrac{(28.3 - 26.7)\,\text{N m}^{-1}}{27.5\,\text{N m}^{-1}} \times 100\% = 6$

 b percentage difference $= \dfrac{(9.81 - 9.71)\,\text{m s}^{-2}}{9.81\,\text{m s}^{-2}} \times 100\% = 1$

56 a $\ln T = \ln\left(\dfrac{2\pi}{\sqrt{k}}\right) + \tfrac{1}{2}\ln m$

b Gradient $= \dfrac{1}{2} = 0.50$

c Intercept $= \ln\left(\dfrac{2\pi}{\sqrt{k/\mathrm{N\,m^{-1}}}}\right) = 0.19$

$\Rightarrow \dfrac{2\pi}{\sqrt{k/\mathrm{N\,m^{-1}}}} = e^{0.19} = 1.21$

$k = \left(\dfrac{2\pi}{1.21}\right)^2 = 27\,\mathrm{N\,m^{-1}}$

57 Using $s = ut + \tfrac{1}{2}at^2$

$1.2\,\mathrm{m} = 0 + \tfrac{1}{2} \times 9.8\,\mathrm{m\,s^{-2}} \times t^2$

$t = 0.49\,\mathrm{s} \approx 0.5\,\mathrm{s}$

58 $R = \dfrac{\rho l}{A} = \dfrac{1.08\times10^{-6}\,\Omega\,\mathrm{m} \times 0.50\,\mathrm{m}}{\pi \times \left(\dfrac{1}{2}\times 0.315\times10^{-3}\,\mathrm{m}\right)^2}$

$R = 6.93\,\Omega \approx 7\,\Omega$

59 $R = \dfrac{\rho l}{A} \Rightarrow A = \dfrac{\rho l}{R} \Rightarrow \dfrac{\pi d^2}{4} = \dfrac{\rho l}{R}$

$\Rightarrow d = \sqrt{\dfrac{4\rho l}{\pi R}}$

$d = \sqrt{\dfrac{4 \times 1.08\times10^{-6}\,\Omega\,\mathrm{m} \times 1.00\,\mathrm{m}}{\pi \times 18.3\,\Omega}}$

$d = 2.74\times10^{-4}\,\mathrm{m}$ (or $0.274\,\mathrm{mm}$)

60 From $V = \varepsilon - Ir$, when $I = 0$ we have voltage $= \varepsilon = 1.58\,\mathrm{V}$. Connecting the resistor across the terminals creates a current I in the cell, so the $1.58\,\mathrm{V}$ drops by Ir to $1.50\,\mathrm{V}$.

$I = \dfrac{\varepsilon}{R + r}$ and $I = \dfrac{V}{R}$

$\Rightarrow \dfrac{\varepsilon}{V} = 1 + \dfrac{r}{R} \Rightarrow r = R \times \left(\dfrac{\varepsilon}{V} - 1\right)$

$r = 4.7\,\Omega \times \left(\dfrac{1.58\,\mathrm{V}}{1.50\,\mathrm{V}} - 1\right) = 0.25\,\Omega$

61 $V = \varepsilon - Ir$

$1.70\,\mathrm{V} = 2.68\,\mathrm{V} - (1.20\times10^{-3}\,\mathrm{A}) \times r$

$r = 817\,\Omega \approx 800\,\Omega$

62

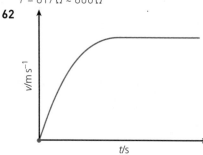

The sphere initially accelerates. When the velocity is such that the viscous drag plus upthrust on the sphere equal the weight of the sphere, the resultant force is zero, and so the velocity becomes constant — the *terminal velocity*.

63 $v = \dfrac{2(\rho_s - \rho_l)\,gr^2}{9\eta}$

$\Rightarrow \eta = \dfrac{2(\rho_s - \rho_l)\,gr^2}{9v}$

Units of $\eta = \dfrac{\mathrm{kg\,m^{-3} \times m\,s^{-2} \times m^2}}{\mathrm{m\,s^{-1}}}$

$= \mathrm{kg\,m^{-1}\,s^{-1}} = \mathrm{kg\,m^{-1}\,s^{-2}} \times \mathrm{s} = \mathrm{Pa\,s}$

64 $m = \rho \times V = \rho \times \tfrac{4}{3}\pi r^3$

$= 7800\,\mathrm{kg\,m^{-3}} \times \tfrac{4}{3}\pi\,(1.5\times10^{-3}\,\mathrm{m})^3$

$= 1.1\times10^{-4}\,\mathrm{kg} \approx 0.1\,\mathrm{g}$

65 $\mathrm{N} = \mathrm{kg\,m\,s^{-2}}$ and $\mathrm{Pa} = \mathrm{N\,m^{-2}} = \mathrm{kg\,m\,s^{-2} \times m^{-2}}$, so $\mathrm{Pa\,s} = \mathrm{kg\,m\,s^{-2} \times m^{-2} \times s} = \mathrm{kg\,m^{-1}\,s^{-1}}$

66 $E = \dfrac{Fl}{A\Delta l} \Rightarrow \Delta l = \dfrac{Fl}{AE}$

$\Delta l = \dfrac{5.0\,\mathrm{N} \times 3.0\,\mathrm{m}}{\pi \times (0.188\times10^{-3}\,\mathrm{m})^2 \times 1.2\times10^{11}\,\mathrm{N\,m^{-2}}}$

$= 1.1\times10^{-3}\,\mathrm{m} \approx 1\,\mathrm{mm}$

67 Stress $= F/A$

$= \dfrac{2.00\,\mathrm{kg} \times 9.8\,\mathrm{N\,kg^{-1}}}{\pi \times (0.19\times10^{-3}\,\mathrm{m})^2} = 1.73\times10^8\,\mathrm{Pa}$

Strain $= \Delta l/l$

$= \dfrac{4.0\times10^{-3}\,\mathrm{m}}{2.94\,\mathrm{m}} = 1.36\times10^{-3}$

$E = \dfrac{\text{stress}}{\text{strain}} = \dfrac{1.73\times10^8\,\mathrm{Pa}}{1.36\times10^{-3}} = 1.3\times10^{11}\,\mathrm{Pa}$

68 Units for E are $\dfrac{\mathrm{N\,m^{-1} \times m}}{\mathrm{m^2}} = \mathrm{N\,m^{-2}} = \mathrm{Pa}$

69 Assume speed of sound $= 330\,\mathrm{m\,s^{-1}}$. Then $f = c/\lambda = 330\,\mathrm{m\,s^{-1}}/0.30\,\mathrm{m} = 1100\,\mathrm{Hz} \approx 1\,\mathrm{kHz}$

70 $T = 4\,\mathrm{div} \times 500\,\mathrm{\mu s\,div^{-1}} = 2000\,\mathrm{\mu s}$ (or $2\,\mathrm{ms}$)

$f = 1/T = 1/(2000\,\mathrm{\mu s}) = 500\,\mathrm{Hz}$

$V = 2.5\,\mathrm{div} \times 500\,\mathrm{mV\,div^{-1}} = 1.25\,\mathrm{V}$

71 $f = \dfrac{1}{2l}\sqrt{\dfrac{T}{\mu}}$

$f = \dfrac{1}{2 \times 0.05\,\mathrm{m}}\sqrt{\dfrac{10\,\mathrm{N}}{2\times10^{-3}\,\mathrm{kg\,m^{-1}}}}$

$= 70.7\,\mathrm{Hz} \approx 70\,\mathrm{Hz}$

72 $\dfrac{1}{\mathrm{N^{-1}\,s^{-2} \times m^2}} = \mathrm{N\,s^2\,m^{-2}}$

$= \mathrm{kg\,m\,s^{-2} \times s^2\,m^{-2}} = \mathrm{kg\,m^{-1}}$

73 $n\lambda = d\sin\theta$

$\Rightarrow \sin\theta = \dfrac{n\lambda}{d} = \dfrac{1 \times 550\times10^{-9}\,\mathrm{m}}{(300\times10^3\,\mathrm{m^{-1}})^{-1}} = 0.165$

$\theta = \sin^{-1}(0.165) = 9.5° \approx 10°$

74 a The distances would be greater (as the wavelength is longer).

  b $n = \dfrac{d\sin\theta}{\lambda}$ and $\sin\theta \le 1$

  $n \le \dfrac{(300 \times 10^3\,\text{m}^{-1})^{-1}}{693 \times 10^9\,\text{m}} \le 4.8$

  n must be an integer, so $n_{max} = 4$

75 $F = 0.040\,\text{kg} \times 9.81\,\text{N kg}^{-1} = 0.392\,\text{N}$
  $m = 0.060\,\text{kg} + 0.440\,\text{kg} = 0.500\,\text{kg}$
  $F = ma \Rightarrow a = F/m$
  $= 0.392\,\text{N}/0.500\,\text{kg}$
  $= 0.78\,\text{N kg}^{-1} \approx 0.8\,\text{m s}^{-2}$

76 In any interaction between bodies, linear momentum is conserved *provided that no resultant external force acts on the bodies.*

77 a $t = RC$
  $t = 3.3 \times 10^3\,\Omega \times 220 \times 10^{-6}\,\text{F} = 0.73\,\text{s} \sim 1\,\text{s}$

  b In the equation $V = V_0 e^{-t/RC}$, t/RC must be just a number, without any unit. Thus RC must have the same unit as t, i.e. second.

  c $R = \dfrac{V}{I}$ and $C = \dfrac{Q}{V}$, so

  $RC = \dfrac{V}{I} \times \dfrac{Q}{V} = \dfrac{Q}{I} = \dfrac{C}{Cs^{-1}} = s$

78 $I = V/R = 5.8\,\text{V}/(3.34 \times 10^3\,\Omega) = 1.7\,\text{mA}$

79 A negative temperature coefficient thermistor is one for which the resistance decreases with an increase in temperature.

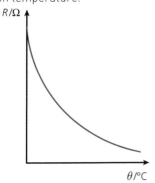

80 $V_{out} = V_{in} \times \dfrac{R_2}{R_1 + R_2}$

  $V_{out} = 6.0\,\text{V} \times \dfrac{330\,\Omega}{(330 + 470)\,\Omega} = 2.48\,\text{V}$

  $= 2.5\,\text{V}$ within the 2% tolerance

81 $\Delta E = L_f \Delta m$
  $= 330\,\text{kJ kg}^{-1} \times 0.010\,\text{kg} = 3.3\,\text{kJ}$

82 Energy cannot be created or destroyed, but is merely changed from one form to another.

83 $\Delta E = mc\Delta\theta$
  $= 0.340\,\text{kg} \times 4200\,\text{J kg}^{-1}\text{K}^{-1} \times (90 - 20)\,\text{K}$
  $= 99960\,\text{J} \approx 0.1\,\text{MJ}$

84 $\Delta E = L_v \Delta m$ and $\Delta E = P\Delta t$, so
  $2.30\,\text{MJ kg}^{-1} \times 0.20\,\text{kg} = 4.6 \times 10^5\,\text{J}$
  $2000\,\text{J s}^{-1} \times t = 4.6 \times 10^5\,\text{J}$
  $\Rightarrow t = 230\,\text{s} \approx 4\,\text{min}$

85

86

  Units of pV are $\text{Pa m}^3 = \text{N m}^{-2}\text{m}^3 = \text{N m} = \text{J}$

87 From $A = A_0 e^{-\mu x}$, a graph of A against x is an *exponential* curve, starting at $A = A_0$ with a downward concave curve that *does not meet the x-axis.* The axes should be labelled A and x.

88 Your graph should have the axes labelled A/min^{-1} and x/mm, with a *smooth* exponential curve of best fit. The half-thickness can be read off for $A = 202.5\,\text{min}^{-1}$, giving a value for $x_{1/2}$ of about 11.6 mm. The activity will fall to a quarter of its initial value in *two* half thicknesses, i.e. 23.2 mm. This shows that the graph is exponential.

89 Squaring $T = 2\pi\sqrt{\dfrac{m}{k}}$ on both sides gives

  $T^2 = \dfrac{4\pi^2 m}{k} = \dfrac{4\pi^2}{k} \times m$

90 From the above answer, a graph of T^2 against m will have a gradient of $4\pi^2/k$. So:

  $\text{gradient} = \dfrac{(0.550 - 0.00)\,\text{s}^2}{(0.325 - 0.00)\,\text{kg}} = 1.69\,\text{s}^2\,\text{kg}^{-1}$

  $k = \dfrac{4\pi^2}{\text{gradient}} = \dfrac{4\pi^2}{1.69\,\text{s}^2\,\text{kg}^{-1}} = 23.3\,\text{kg s}^{-2}$

Index

Index